Mathematical Modeling in Biology

Mathematical Modeling in Biology: A Research Methods Approach is a textbook written primarily for advanced mathematics and science undergraduate students and graduate-level biology students. Although the applications center on ecology, the expertise of the authors, the methodology can be imported to any other science, including social science and economics. The aim of this book, beyond being a useful aid to teaching and learning the core modeling skills needed for mathematical biology, is to encourage students to think deeply and clearly about the meaning of mathematics in science and to learn significant research methods. Most importantly, it is hoped that students will experience some of the excitement of doing research.

Features

- Minimal prerequisties are a solid background in calculus, such as a Calculus I course

- Suitable for upper division mathematics and science students and graduate-level biology students

- Provides sample MATLAB codes and instruction in appendices along with data sets available on https://bit.ly/3fcLF3D.

Chapman & Hall/CRC Mathematical Biology Series

Series Editors:
Ruth Baker, Mark Broom, Adam Kleczkowski, Doron Levy,
Sergei Petrovskiy

About the Series

This series aims to capture new developments in mathematical biology, as well as high-quality work summarizing or contributing to more established topics. Publishing a broad range of textbooks, reference works, and handbooks, this series is designed to appeal to students, researchers, and professionals in mathematical biology.

We will consider proposals on all topics and applications within the field, including but not limited to stochastic modeling, differential equation modeling, dynamical systems, game theory, machine learning, data science, evolutionary biology, cell biology, oncology, epidemiology, and ecology.

Systems Biology
Mathematical Modeling and Model Analysis
Andreas Kremling

Cellular Potts Models
Multiscale Extensions and Biological Applications
Marco Scianna, Luigi Preziosi

Quantitative Biology
From Molecular to Cellular Systems
Edited By Michael E. Wall

Game-Theoretical Models in Biology, Second Edition
Mark Broom, Jan Rychtář

Mathematical Modeling in Biology
A Research Methods Approach
Shandelle M. Henson, James L. Hayward

For more information about the series, visit:
https://www.routledge.com/Chapman–HallCRC-Mathematical-Biology-Series/book-series/CRCMBS

Mathematical Modeling in Biology

A Research Methods Approach

Shandelle M. Henson and James L. Hayward

CRC Press
Taylor & Francis Group
Boca Raton London New York

CRC Press is an imprint of the
Taylor & Francis Group, an **informa** business

A CHAPMAN & HALL BOOK

Cover photo © by James L. Hayward

First edition published 2023
by CRC Press
4 Park Square, Milton Park, Abingdon, Oxon, OX14 4RN

and by CRC Press
6000 Broken Sound Parkway NW, Suite 300, Boca Raton, FL 33487-2742

British Library Cataloguing-in-Publication Data
A catalogue record for this book is available from the British Library

Library of Congress Cataloging-in-Publication Data
Names: Henson, Shandelle Marie, 1964-author. | Hayward, James L., 1948-author.
Title: Mathematical modeling in biology : a research methods approach /
Shandelle M. Henson, Andrews University, USA, James L. Hayward, Andrews
University, USA.
Description: First edition. | Boca Raton : C&H/CRC Press, 2023. |
Series: Chapman and Hall/CRC mathematical biology series |
Includes bibliographical references and index.
Identifiers: LCCN 2022030360 (print) | LCCN 2022030361 (ebook) |
ISBN 9781032208213 (hardback) | ISBN 9781032206943 (paperback) |
ISBN 9781003265382 (ebook)
Subjects: LCSH: Biology–Mathematical models.
Classification: LCC QH323.5 .H46 2023 (print) | LCC QH323.5 (ebook) |
DDC 570.1/5118–dc23/eng/20220728
LC record available at https://lccn.loc.gov/2022030360
LC ebook record available at https://lccn.loc.gov/2022030361

ISBN: 9781032208213 (hbk)
ISBN: 9781032206943 (pbk)
ISBN: 9781003265382 (ebk)

DOI: 10.1201/9781003265382

Typeset in Minion
by codeMantra

To our parents
Audrey Gackenheimer Henson and
John William Henson III
Jane Watson Hayward and James Lloyd Hayward, Sr.

Contents

ACKNOWLEDGMENTS   xvii

FOR PROFESSORS AND STUDENTS   xix

AUTHORS   xxi

SECTION I   **Introduction to Modeling**

CHAPTER 1 ▪ Mathematical Modeling   3

1.1   WHAT YOU SHOULD KNOW ABOUT THIS CHAPTER   3

1.2   THE MODELING CYCLE   4

    1.2.1   Step 1: Translation into Mathematics   5

        1.2.1.1   Choosing Variables and Parameters   5

        1.2.1.2   Simplifying Assumptions   6

        1.2.1.3   Parameterization   7

    1.2.2   Step 2: Model Analysis   7

    1.2.3   Step 3: Back-Translation   7

        1.2.3.1   Model Selection and Validation   8

        1.2.3.2   Test of Model Predictions   8

    1.2.4   Step 4: Revising Model Assumptions   8

1.3   BIOLOGY   9

    1.3.1   Ecology   10

1.4   MATHEMATICS   13

1.5   STATISTICS   16

1.6   EPISTEMOLOGY: HOW WE KNOW   19

1.7   EXERCISES   23

BIBLIOGRAPHY   28

CHAPTER 2 ▪ Avian Bone Growth: A Case Study   29

2.1   WHAT YOU SHOULD KNOW ABOUT THIS CHAPTER   29
2.2   SCIENTIFIC PROBLEM   30
      2.2.1   Data   30
2.3   TRANSLATION INTO MATHEMATICS   32
      2.3.1   Simplifying Assumptions   32
              2.3.1.1   Deterministic Assumptions   33
              2.3.1.2   Stochastic Assumptions   33
      2.3.2   The Deterministic Model   33
      2.3.3   The Stochastic Model   34
2.4   MODEL PARAMETERIZATION   36
      2.4.1   Dividing the Data Set   36
      2.4.2   Maximum Likelihood (ML) Method   38
      2.4.3   Nonlinear Least Squares (LS) Method   39
      2.4.4   Downhill Minimization Routine:
              Nelder-Mead Algorithm   40
      2.4.5   Implementing Parameterization in Code   41
      2.4.6   Results of Parameterization   42
2.5   MODEL SELECTION   42
2.6   MODEL VALIDATION   45
2.7   EXERCISES   46
      BIBLIOGRAPHY   51

SECTION II Discrete-Time Models

CHAPTER 3 ▪ Discrete-Time Maps   55

3.1   WHAT YOU SHOULD KNOW ABOUT THIS CHAPTER   55
3.2   COMPARTMENTAL MODELS   55
3.3   LINEAR MAPS   57
      3.3.1   Malthusian Growth   57
3.4   NONLINEAR MAPS   60

3.5	LINEARIZATION	62
	3.5.1 Linearization of Functions	62
	3.5.2 Linearization of Discrete-Time Maps	64
	3.5.3 Linearizing the Ricker Map	67
3.6	THE RICKER NONLINEARITY	70
3.7	EXERCISES	71
	BIBLIOGRAPHY	74

CHAPTER 4 ■ Chaos: Simple Rules Can Generate Complex Results 75

4.1	WHAT YOU SHOULD KNOW ABOUT THIS CHAPTER	75
4.2	RICKER MODEL REVISITED	75
4.3	NEW PARADIGMS ARISE FROM CHAOS	80
	4.3.1 Deterministic Unpredictability	80
	4.3.2 Complex Results Can Arise from Simple Rules	80
4.4	MAY'S HYPOTHESIS	81
4.5	EXERCISES	81
	BIBLIOGRAPHY	84

CHAPTER 5 ■ Higher-Dimensional Discrete-Time Models 85

5.1	WHAT YOU SHOULD KNOW ABOUT THIS CHAPTER	85
5.2	INTRASPECIFIC INTERACTIONS	86
5.3	INTERSPECIFIC INTERACTIONS	86
5.4	EXAMPLE OF AN AGE-STRUCTURED SINGLE-SPECIES MODEL	87
5.5	EXAMPLE OF A TWO-SPECIES MODEL	89
5.6	n-DIMENSIONAL LINEAR DIFFERENCE EQUATIONS	90
	5.6.1 n-Dimensional Leslie Models	90
5.7	SOLVING LINEAR SYSTEMS OF DIFFERENCE EQUATIONS	92
	5.7.1 An Example	92

5.7.2   Solving the General Two-Dimensional System   96

5.7.3   Solving Higher-Dimensional Systems   97

5.8   NONLINEAR SYSTEMS   99

5.8.1   Linearization   100

5.8.2   An Example   102

5.9   EXERCISES   104

BIBLIOGRAPHY   109

CHAPTER   6 ▪ Flour Beetle Dynamics: A Case Study   111

6.1   WHAT YOU SHOULD KNOW ABOUT THIS CHAPTER   111

6.2   FLOUR BEETLES   111

6.3   DATA   113

6.4   ASSUMPTIONS   114

6.4.1   Deterministic Assumptions   114

6.4.2   Stochastic Assumptions   115

6.5   ALTERNATIVE DETERMINISTIC MODELS   116

6.6   STOCHASTIC MODELS   117

6.7   MODEL PARAMETERIZATION   119

6.7.1   Conditioned One-Step Residuals   119

6.7.2   Conditioned Least Squares (CLS)   120

6.8   MODEL SELECTION   122

6.9   MODEL VALIDATION   124

6.10   THE "HUNT FOR CHAOS" EXPERIMENTS   125

6.10.1   Results of the "Hunt for Chaos" Experiments   128

6.10.2   Manipulating the Parameters in the
Laboratory   128

6.11   EXERCISES   130

BIBLIOGRAPHY   136

SECTION III Continuous-Time Models

CHAPTER 7 ▪ Introduction to Differential Equations 145

7.1 WHAT YOU SHOULD KNOW ABOUT THIS CHAPTER 145
7.2 COMPARTMENTAL MODELS 146
    7.2.1 A Tank Problem 146
    7.2.2 The SIR Model 147
    7.2.3 The Continuous-Time Logistic Model 150
7.3 EXERCISES 154
BIBLIOGRAPHY 156

CHAPTER 8 ▪ Scalar Differential Equations 159

8.1 WHAT YOU SHOULD KNOW ABOUT THIS CHAPTER 159
8.2 LINEAR EQUATIONS 159
    8.2.1 Malthusian Growth 161
8.3 NONLINEAR EQUATIONS 162
    8.3.1 Logistic Growth 162
    8.3.2 Allee Effects 164
    8.3.3 The "Doomsday Model" of Human Population Growth 166
8.4 LINEARIZATION 170
8.5 BIFURCATIONS 175
    8.5.1 Transcritical Bifurcation 176
    8.5.2 Saddle-Node Bifurcation 177
    8.5.3 Pitchfork Bifurcation 177
    8.5.4 Hysteresis 178
8.6 EXERCISES 179
BIBLIOGRAPHY 185

CHAPTER 9 ▪ Systems of Differential Equations 187

9.1 WHAT YOU SHOULD KNOW ABOUT THIS CHAPTER 187

9.2   LINEAR SYSTEMS OF ODES AND PHASE PLANE ANALYSIS 188

    9.2.1   Unstable Node   189

    9.2.2   Asymptotically Stable Node   191

    9.2.3   Saddle   192

    9.2.4   Center   193

    9.2.5   Unstable Spiral   195

    9.2.6   Asymptotically Stable Spiral   196

    9.2.7   Summary: Eigenvalues Tell All   197

9.3   NONLINEAR SYSTEMS OF ODES   198

    9.3.1   Linearization   201

9.4   LIMIT CYCLES, CYCLE CHAINS, AND BIFURCATIONS   202

9.5   LOTKA-VOLTERRA MODELS AND NULLCLINE ANALYSIS 204

    9.5.1   Lotka-Volterra Competition   205

    9.5.2   Lotka-Volterra Cooperation   205

    9.5.3   Lotka-Volterra Predator-Prey   206

9.6   EXERCISES   208

BIBLIOGRAPHY   213

CHAPTER 10 ▪ Seabird Behavior: A Case Study   215

10.1   WHAT YOU SHOULD KNOW ABOUT THIS CHAPTER   215

10.2   THE SCIENTIFIC PROBLEM   215

10.3   HISTORICAL DATA   217

    10.3.1   Count Data   217

    10.3.2   Dividing the Data   217

    10.3.3   Tide and Solar Elevation Data   217

10.4   GENERAL MODEL   218

10.5   ALTERNATIVE MODELS   220

10.6   MODEL PARAMETERIZATION   220

10.7   MODEL SELECTION   221

10.8   MODEL VALIDATION   222

10.9   TEST OF A PRIORI PREDICTIONS   222

10.10   STEADY-STATE MODEL   226

10.11   DISCUSSION   227

10.11.1	Importance of Scale	227
10.11.2	Resource Management	228
10.12	EXERCISES	228
	BIBLIOGRAPHY	230

SECTION IV Regression Models

CHAPTER 11 ▪ Introduction to Regression 237

11.1	WHAT YOU SHOULD KNOW ABOUT THIS CHAPTER	237
11.2	LINEAR REGRESSION	237
11.2.1	Simple Linear Regression (Single Factor)	238
11.2.2	Multiple Linear Regression (Multiple Factors)	238
11.2.3	Stochastic Model and Parameter Estimation	239
11.2.4	Confidence Intervals for Regression Coefficients	240
11.3	LOGISTIC REGRESSION	240
11.3.1	Odds Ratios (ORs)	241
11.3.2	OR Confidence Intervals	242
11.4	GENERALIZED LINEAR MODELS (GLMs)	242
11.5	INTERACTION TERMS	242
11.6	EXERCISES	243
	BIBLIOGRAPHY	246

CHAPTER 12 ▪ Climate Change and Seabird Cannibalism: A Case Study 247

12.1	WHAT YOU SHOULD KNOW ABOUT THIS CHAPTER	247
12.2	THE SCIENTIFIC PROBLEM	248
12.3	DATA	251
12.4	LOGISTIC REGRESSION ANALYSIS	252
12.5	MODEL VALIDATION	253
12.6	OUTCOMES	253
12.7	CLIMATE CHANGE, CANNIBALISM, AND REPRODUCTIVE SYNCHRONY	255

12.8   EXERCISES   257

BIBLIOGRAPHY   259

Section V Appendix

Appendix A ▪ Linear Algebra Basics   265

A.1   MATRIX OPERATIONS   265

   A.1.1   Matrix Addition   265

   A.1.2   Scalar Multiplication   265

   A.1.3   Matrix Subtraction   266

   A.1.4   Matrix Multiplication   266

   A.1.5   Determinants of Square Matrices   267

A.2   EXERCISES   267

A.3   SOLUTIONS   270

A.4   SUMMARY OF LINEAR ALGEBRA CONCEPTS   272

Appendix B ▪ MATLAB: The Basics   273

B.1   PRELIMINARIES   273

B.2   SYNTAX AND PROGRAMMING   273

   B.2.1   Command Line   273

   B.2.2   Case-Sensitivity   274

   B.2.3   Displaying the Current Value of a Variable   274

   B.2.4   Clearing All Variables   274

   B.2.5   Closing MATLAB   274

   B.2.6   Variables and Arithmetic Operators   274

   B.2.7   Programs   276

   B.2.8   Comment Lines   277

   B.2.9   Printing to the Screen   277

   B.2.10   Numerical Format   278

   B.2.11   Loops   278

   B.2.12   Crashing a Program on Purpose   279

B.2.13   Logical Statements (If-Then-Else)   279

B.2.14   Input and Output Files   280

B.2.15   Creating Functions   281

B.2.16   Subroutines   282

B.2.17   Vectors, Matrices, and Arrays   283

B.2.18   Functions in the MATLAB Library   285

B.2.19   Plotting   286

B.2.20   Simulating Discrete-Time Models   286

B.2.21   Simulating Ordinary Differential
         Equations (ODEs)   290

B.2.22   The Downhill Nelder-Mead Algorithm   292

B.3   EXERCISES   293

APPENDIX C ▪ Connecting Models to Data: A Brief Summary
          with Sample Codes   295

C.1   PARAMETERIZATION   295

C.2   RESIDUAL ERRORS (RESIDUALS)   296

C.3   RSS AND R^2 296

C.4   MAXIMUM LIKELIHOOD (ML) PARAMETERS   297

C.5   LEAST SQUARES (LS) PARAMETERS   297

C.6   IMPLEMENTATION IN CODE   298

C.6.1   Basic Structure of Program   298

C.6.2   Constructing Input Files   299

C.6.3   Example: Algebraic Model Using Vectors   299

C.6.4   Example: Algebraic Model Using Loop   300

C.6.5   Example: Scalar Map with One-Step
        Predictions   302

C.6.6   Example: Higher-Dimensional Discrete-Time
        Model with One-Step Predictions   303

Index   307

Acknowledgments

We thank the U. S. National Science Foundation for funding the research of the Beetle Team and Seabird Ecology Team, whose data and models form the case studies in this book. We also thank the Office of Research and Creative Scholarship at Andrews University for funding our research and students. Members of the Beetle Team have been especially influential in the preparation of this book, namely Robert F. Costantino, Jim M. Cushing, Brian Dennis, Robert A. Desharnais, and Aaron A. King, as well as senior members of the Seabird Ecology Team Gordon J. Atkins, Jim M. Cushing (again), Joseph G. Galusha, and Lynelle M. Weldon. We also thank the co-authors of articles on which the case study chapters are based. Our research students and the students who have taken the mathematical modeling class deserve much credit for inspiring and testing this book. We express special thanks to our Ph.D. advisors, Thomas G. Hallam and Don E. Miller, and to S.M.H.'s postdoc advisor Jim M. Cushing (again!), who turned us into professionals and sharpened our thinking. Several institutions have supported our work logistically, including Rosario Beach Marine Laboratory, U. S. Fish and Wildlife Service, and Cape George Colony which provided use of its marina for access to Protection Island. We are particularly indebted to Ross Anderson of Cape George for his friendship and numerous kindnesses during our research. It was a joy to work with Jessica Rim, who prepared the figures. Finally, we appreciate the kind editors at CRC Press/Taylor & Francis Group.

For professors and students

This textbook has been inspired by our graduate and undergraduate research students and the fascinating work with which they have been engaged over the last few decades. It is a research methods textbook for a mathematical modeling course, written for upper division mathematics and science students and graduate-level biology students. Although the methodology in this text is presented in the context of ecology because that is the research area of the authors, it can be imported to any other science, including social science and economics.

The prerequisite for this textbook is a solid background in differential calculus and an introduction to integral calculus at least through the method of integration by substitution. This prerequisite is equivalent to most one-semester university Calculus I courses. The text contains three appendices that easily lead the student through series of exercises to develop all other necessary skills from linear algebra and computer programming. Although the programming language is not specified in the text—an instructor can use any scientific programming language—the sample codes and instruction in the appendices use MATLAB. Thus, if students are first-time programmers, the easiest approach for both teacher and student would be to use MATLAB, unless the instructor wants to rewrite the sample instructional codes in another language.

Our intent in this book is to integrate mathematical theory, modeling, real data from the literature, and programming in such a way that students gain significant research skills. The "Case Study" chapters involve connecting models to data. Most of the exercises are not "routine" but rather require students to engage in a way that a researcher would, bringing together many skills to solve novel problems. Although there is not a large number of exercises, they are substantial. Many of them can be assigned as projects. Even though Calculus I is the only prerequisite, this approach requires a level of intellectual maturity in students. Only students who are ready to engage in open-ended and sometimes frustrating problem solving leading to unknown outcomes should take the course. At our university, we have used the text for a research methods "swing course"—a capstone course for undergraduate mathematics and science majors and a beginning course for biology graduate students. Highly motivated, curious, research-oriented freshmen and sophomores will be successful, as well.

Each chapter has an annotated bibliography. In general we have restricted the references to classic or foundational papers. In the "Case Study" chapters, we also include the body of literature on which the case study is based. The short annotations below most bibliography entries allow the reader to understand the relevance at a glance.

We hope this text will help students think deeply and clearly about the meaning of mathematics in science and learn some of the research methods of applied mathematics. Most importantly, we hope that students will experience some of the excitement of doing research. Researchers are explorers who have a passion to find and understand the deep structure of the universe. May each student find a passion for learning and discovery.

Authors

Shandelle M. Henson is Professor of Mathematics and Professor of Ecology at Andrews University, Michigan, the USA. She uses dynamical systems theory and bifurcation theory to study nonlinear population dynamics and the effects of climate change on marine organisms. Shandelle earned a Ph.D. in mathematics at the University of Tennessee, Knoxville, and did several years of postdoctoral work at the University of Arizona. She serves as Editor-in-Chief of the journal *Natural Resource Modeling* and is a co-author of the book *Chaos in Ecology: Experimental Nonlinear Dynamics* (Academic Press 2003), which documented chaotic dynamics in laboratory populations of insects.

James L. Hayward is Professor Emeritus of Biology at Andrews University, Michigan, the USA. He earned a Ph.D. in zoology from Washington State University, and he has studied the behaviors and population dynamics of marine birds and mammals in the Pacific Northwest of North America since beginning graduate studies in 1972. In addition, Jim has studied the behavior of marine iguanas and flightless cormorants on the remote and uninhabited island of Fernandina, the westernmost island in the Galápagos. He also is a recognized expert in the fossilization of eggshell in birds and dinosaurs.

Shandelle and Jim, a wife-husband research team, are widely published in both the technical and popular literature, and both have won awards for their teaching. They have applied mathematical methods to the behavioral dynamics of seabirds at Protection Island National Wildlife Refuge in the Strait of Juan de Fuca, Washington, the USA, since 2002. They reside in Niles, Michigan, the USA, and they have a daughter, son-in-law, and four grandsons. They enjoy hiking, geology, art, music, and literature.

I

Introduction to Modeling

CHAPTER 1

Mathematical Modeling

1.1 WHAT YOU SHOULD KNOW ABOUT THIS CHAPTER

We begin this textbook with an overview of the concept of mathematical modeling. What is mathematical modeling, how is it done, and how does it fit into the scientific method? How do the tools of mathematics and science together fit into the larger context of the humanities?

Mathematical models may be "proof-of-concept," or quite realistic or somewhere in between. A proof-of-concept model is a toy model, not tied rigorously to data from a specific system, that incorporates a mechanism(s) of interest. The mechanism is mathematically probed and perturbed by the modeler in order to determine implications for the outcomes. Mathematical results from such a model analysis suggest hypotheses for the real-world system of interest. A realistic model, on the other hand, is tied rigorously to data from a particular system. Of course, even a realistic model has simplifications by design and is not as complicated as the original system. In general, the more realistic a mathematical model is, the less tractable it is. The "art of modeling" is the skill of finding that middle ground in which the model is tractable, but realistic enough to be informative.

In addition, mathematical models may be phenomenological or mechanistic. Phenomenological models describe the observed systems via curve fitting, but are not based on first principles. The equations of mechanistic models, on the other hand, are based on the main mechanisms thought to drive the dynamics of the system. The most

DOI: 10.1201/9781003265382-2

powerful mathematical models not only describe, but also explain and predict real-world systems. In general, models are most useful scientifically if their formulations are mechanistic.

With these general ideas in mind, let's take a closer look at mathematical models.

1.2 THE MODELING CYCLE

A *mathematical model* is an equation or system of equations that describes a physical or biological system. For example, the algebraic model

$$v(t) = -gt \tag{1.1}$$

is an equation that describes the velocity of an object dropped from rest near the surface of a planet, where $g > 0$ is the acceleration due to gravity and t is time.

Models often describe *dynamical systems*, that is, systems that change through time. Since continuous-time rates of change are derivatives, models often take the form of *differential equations* (equations that involve derivatives). For example, the algebraic model (1.1) can be written as the differential equation model

$$\frac{ds}{dt} = -gt, \tag{1.2}$$

where $s(t)$ is the distance traveled by the object. The differential equation model (1.2) is a *continuous-time model*.

Some physical and biological processes are better described by *discrete-time models*. For example, if each individual in a population of annual plants produces b seeds per year, a fraction p of which germinate, then the number of individual plants the next year is

$$N_{t+1} = pbN_t, \tag{1.3}$$

where N_t is the number of plants in the population at year t and the parameters satisfy $b > 0$ and $0 < p < 1$. Model (1.3) is called a *difference equation*, a *discrete-time map*, or a *recursion formula*.

Mathematical models are effective tools for addressing scientific problems. This is because a mathematical model is the translation of a scientific problem into the precise language of mathematics,

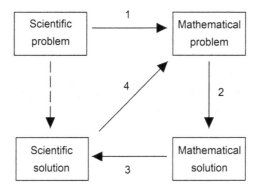

Figure 1.1: The process of mathematical modeling.

where it becomes amenable to many powerful techniques. Figure 1.1 shows a conceptual diagram for the process of mathematical modeling. Important issues arise in each of the four modeling steps.

1.2.1 Step 1: Translation into Mathematics

The first step in modeling is to translate the scientific problem into the precise language of mathematics. This process requires a crisp conceptual clarification that often proves valuable in and of itself.

1.2.1.1 Choosing Variables and Parameters

We must decide what quantitative aspects of the system we wish to follow and name those quantities with variables. Input variables are called *independent variables*. Time, location in space, and age are examples of common independent variables. *Dependent variables* are the functions of interest that depend on the independent variables. In modeling, dependent variables often are called *state variables*, because they track the state of the system. For example, suppose we want to study how a population is distributed in time and space. We could take $N(t, x)$ to be the density of organisms at time t at location x. Here, t and x are the independent variables and N is the state variable. In equation (1.2), the state variable is s and the independent variable is t. We know this because the presence of the derivative ds/dt alerts us to the fact that $s(t)$ is the function under consideration.

Besides the independent variables, the system in question may depend on other *extrinsic* (environmental) factors that are relatively constant on the time scale of the problem. Typical examples are temperature and humidity. We might include these factors in the equation as constants. Such constants are called *parameters*. (Some people call them "model coefficients," but this is not the best terminology because the constants do not always appear in the equation as true coefficients; for example, they may appear as terms or as exponents.) In equation (1.1), t is the independent variable, v is the state variable, and g is an extrinsic parameter (the acceleration due to gravity). Earth has one value of g (about 9.8 m/s^2), while Mars has another (about 3.7 m/s^2). Other parameters may be *intrinsic*. In equation (1.3), N is the state variable, t is the independent variable, and b is an intrinsic parameter. The parameter p might be a compendium of both extrinsic and intrinsic factors.

1.2.1.2 Simplifying Assumptions

Once the variables have been identified, they must be linked together with equations. Constructing model equations always involves making simplifying assumptions about the real system; for example, the object is close to the surface of the planet; the force of gravity at the surface of the planet is constant; gravity is the only force acting on the object (no friction). Mathematical models do not perfectly describe the "real world"; they are simplifications that capture the main mechanisms driving the observed patterns. A major goal of modeling and the art of modeling is to construct a simple, mechanistic, low-dimensional (few state variables) model that describes enough of the observed variability so that the leftover, "higher-order" variability is relatively small and can be described statistically as *noise* (random effects). We can add more and more mathematical complications (for example, the force due to friction on the falling object) to a model in an attempt to make it more realistic, but at some point the model becomes unwieldly and intractable. There is little advantage in replacing a real system that is too complicated to understand with a mathematical model that is too complicated to understand. An important modeling mantra is therefore KISS (keep it simple stupid). On the other hand, if the assumptions oversimplify the problem, the mathematical model will not be able to approximate the real system. Thus, in modeling there is always a trade-off between realism and tractability. The goal is to capture the essential behavior of a complicated real system with a simplified mathematical system; the "art" consists in determining the main mechanisms driving the system, building the assumptions around those mechanisms, and consigning lesser influences to noise.

1.2.1.3 Parameterization

If model equations are to describe a real system, they must be connected to real data through *parameterization*. Parameterization is the statistical process by which data are used to estimate the values of the parameters. For example, in equation (1.1) the value of g can be estimated from experimental data on falling objects. You may have done this experiment in a general physics lab. Parameterization is also known as *model fitting* or *parameter estimation*. The word "estimation" in this context is a technical term from statistics that refers to the result of rigorous model fitting.

The parameterized model equation(s) may now be taken, tentatively, as a surrogate for the real system. In fact, the model constitutes a hypothesis about the scientific problem. More than one model can be constructed, based on different hypotheses about what drives the system. These competing models serve as *alternative hypotheses*, which are then tested in Step 3.

1.2.2   Step 2: Model Analysis

The second step in modeling is mathematical analysis. All the powerful concepts and tools of mathematics may now be brought to bear on the scientific problem through its surrogate, the model. The analysis may require solving differential equations, drawing bifurcation diagrams, linearizing, or utilizing any of a host of other techniques you will learn in this book or have learned in other mathematics classes.

1.2.3   Step 3: Back-Translation

The third step is to translate the mathematical results back into the language of the original scientific problem. At this point, we must ask whether the results correspond to extant data, and whether the model can accurately predict new data. If not, then we will need to revise the model and "go around the loop again" (Figure 1.1).

1.2.3.1 Model Selection and Validation

Model selection is the process by which alternative parameterized models are compared to each other in an effort to select the best model. In Chapter 2, we will discuss a statistical tool called the *Akaike information criterion* (AIC), one of the tools from *information theory* that researchers often use for model selection. *Model validation* is the process of evaluating a model by comparing its output to new data. If there is a pronounced lack of correspondence with data, then the modeling assumptions need to be revised. *Model validation should be a procedure distinct from model fitting.* Ideally, models should be validated on independent data sets, that is, on data sets that were not used to estimate the model parameters. Model validation involves computing the so-called *goodness-of-fit* statistics that measure how closely model output corresponds to data.

1.2.3.2 Test of Model Predictions

A good model not only describes and explains, but also predicts. The best way to further test a validated model is to make *a priori* predictions and then collect data to test those predictions. Ideally, the predictions should be unusual or unexpected. A successful test of such predictions makes a strong case that the model really does capture the essential behavior of the system.

1.2.4 Step 4: Revising Model Assumptions

The fourth step is ubiquitous in modeling. A model that doesn't correspond to data must be revised. One or more of the assumptions were false or were oversimplifications; important factors may have been left out of the model. In practice, model revision occurs more or less continually throughout the modeling process as new insights are gained and equations are tweaked.

You can see that modeling in biology requires a thorough integration of biology, mathematics, and statistics. The next three sections contain some important general ideas and terms from each of these disciplines.

1.3   BIOLOGY

Biology is the study of living systems. Mathematical models can describe living systems at subcellular, cellular, tissue, organ, organ system, organism, population, community, and ecosystem levels of organization. The level of organization chosen depends on the interest of the investigator and the questions asked. For example, if an epidemiologist (biologist who studies patterns of disease) wants to know how fast a disease will spread through a region, the modeling process would focus on the population inhabiting the region. Modeling subcellular processes might be of little help. In other words, before modeling begins, one must select an appropriate *scale* of interest (Levin 1992).

Living systems are the most complex systems known. In order to model a living system at any scale, we must observe the system thoroughly. In Chapter 10, we explain how we developed a model to accurately predict the number of seabirds occupying a particular habitat at a particular time. Our efforts were successful because we and others spent a lot of time learning about the everyday lives of these birds. Armed with this knowledge, we could ask the right questions, identify the independent variables most likely to influence seabird behavior, and develop an efficient data collection protocol suitable for model parameterization and testing. There is no substitute for an intimate observational knowledge of a living system whose mechanisms you want to understand better through modeling. This is why an applied mathematician who wants to model a biological system should collaborate with a biologist who knows the system well.

Some biologists assume that living systems are too complicated to be modeled mathematically. It may be that at some scales, a particular living system is too complicated to model. It might not be feasible, for example, to model how each of the millions of neurons in your cerebral cortex are working in concert to process information conveyed by the words in this paragraph. A neurobiologist, however, might be able to model the activity of two or three interacting neurons, or entire sections of the brain. Mathematical modeling may not be able to answer all the interesting questions a biologist might ask.

Evolutionary biologists and population geneticists have long taken advantage of mathematical modeling. This is because people like G. H. Hardy, Sir Ronald Fisher, Sewell Wright, and J. B. S. Haldane, trained in mathematics, turned their attention to solving biological problems. Biologists in other subdisciplines have also discovered the power of modeling.

At the most basic level, mathematical modeling provides biologists with a way to qualitatively and quantitatively *describe* the behavior of living systems. At a higher level, modeling provides a tool to *identify* factors that drive living systems. Finally, once these factors have been identified, biologists can *test predictions* about how these systems will function in the future.

Our interest and experience concern *ecological modeling*. Because many of the examples used in this book come from ecology, we now provide a brief description of this subdiscipline.

1.3.1 Ecology

Ecologists are concerned with interactions among organisms and their environments. Both organisms and environments are very complex; moreover, they constantly undergo change. Ecologists thus face significant challenges as they try to uncover ecological patterns.

An interesting challenge that faces all ecologists is the problem of scale. Patterns apparent at one scale may disappear at higher or lower levels of organization, and new patterns may emerge at these levels. Patterns at some scales exhibit *emergent properties*. This is another way of saying the whole often seems more than the sum of the parts— qualities emerge at higher levels that might not be predicted on the basis of the known properties of lower levels. Emergent properties are not magical, however. They are the result of the properties of entities at smaller scales and all of their interactions at the larger scale.

Ecologists view ecosystems as consisting of progressively more inclusive scales:

Individual organism: Individuals live more or less autonomous lives. Members of *solitary species* interact little with conspecifics, whereas members of *social species* interact frequently with one another. Moreover, individuals are sometimes difficult to distinguish. Examples include the assemblage of medusae and polyps in a Portuguese man-o'-war, the collection of individuals forming a sphere of *Volvox*, and a grove of aspen trees generated through vegetative reproduction.

Population: Members of a single species that inhabit a particular geographic region.

Species: A group of interbreeding or potentially interbreeding organisms genetically isolated from other such groups. (Note: This is only one of many competing species concepts. Species concepts generate lots of debate among biologists.)

Community: An assemblage of populations that occupy the same geographic region.

Ecosystem: A community plus all the non-living components of the system. These non-living (*abiotic*) components include things such as energy, moisture, and nutrients.

Ecology includes a number of subdisciplines:

Physiological ecologists (or *environmental physiologists*) deal with how individual organisms cope with environmental change. For example, they have investigated how camels conserve water in the heat of the desert, and how seals and porpoises conserve oxygen during long underwater dives.

Population ecologists are interested in the factors that influence population growth.

Community ecologists are concerned with interactions between and among species that live in the same community. These interactions include food web relations, predator-prey interactions, interspecific competition, mutualism, parasitism, speciation, and coevolution.

Ecosystem ecologists examine how energy flows through and how nutrients cycle within ecosystems. Thus, ecosystem ecologists work with energy budgets and accounting schemes by which the precise nature of community interactions can be teased apart, and they determine pathways taken by nutrients through ecosystems.

Behavioral ecologists study the causation, function, ontogeny, and evolution of the behaviors of animals in their ecological setting.

Mathematical ecologists use theoretical methods such as mathematical models and computational simulations to study ecosystems. They try to understand how the dynamics of populations and ecosystems depend on mechanistic processes as well as biotic and abiotic factors.

Considerable overlap occurs among the areas of ecology. For example, a population ecologist needs to understand the life history characteristics of individual organisms in the population as well as how community interactions such as competition and predator-prey relations impact the population.

Ecologists, like other biologists, are often tempted to confuse *correlation* with *causation*. The fact that we can mathematically model a relationship between particular independent and dependent variables is no proof that the variables are causally linked. Mathematical implication is not the same as causation. For example, we might develop a model based on temperature change that quite accurately predicts when white-crowned sparrows migrate south for the winter. Yet we know from experimental studies that photoperiod, not temperature, is the environmental factor that causes white-crowned sparrows to migrate. Because temperature covaries with the light-dark cycle, we see how easy it would be to arrive at an invalid conclusion concerning the cause of migration. Thus, if we are not careful, mathematical modeling can sometimes lead us to incorrect conclusions. On the other hand, mathematical modeling can often suggest possible causal relationships that we might not have guessed otherwise.

Another point worth making in this context is the difference between *proximate* and *ultimate causes*. In the example just given, the proximate cause of migration is photoperiod change, which alters the neuroendocrine system of white-crowned sparrows. The ultimate cause in this example was natural selection which, over generations, favored the fitness of birds that nested where food was abundant during the spring and moved to warmer regions when nesting areas turned frigid.

1.4 MATHEMATICS

The most basic mathematical concept in modeling is that of the *function*. A relationship between an input t and an output $x(t)$ is called a function if and only if the value of the output $x(t)$ is completely determined by the input t. Symbolically, $x(t)$ is a function if and only if

$$t_1 = t_2 \implies x(t_1) = x(t_2),$$

where the double arrow means "implies." In words, if the inputs are the same, then the outputs must be the same. Said another way, you can't get two different outputs from the same input. Algebra students learn this as the "vertical line test." Functions are *deterministic*. (By contrast, a relation in which you *can* get two different outputs from the same input is called *stochastic*; we will turn our attention to stochasticity in the next section.)

In modeling dynamical systems, we are interested in the state of the system (call it x) as a function of time t. That is, we are interested in the behavior of $x(t)$ as t changes. We can list the corresponding values of x and t in a table, and we can also graph x vs. t. This list of pairs of numbers (or its graph) is called a *time series*. You will see many times series graphs in later chapters (e.g., Figures 4.2 and 8.2). If there are two state variables for the system of interest, say $x(t)$ and $y(t)$, we must show two time series graphs together: x vs. t and y vs. t.

In general, the two state variables x and y may be *coupled*, that is, may depend on each other, so the two time series graphs must be interpreted together. This is a visually difficult task. A better visual tool in this case is that of *state space*, or *phase space,* in which the state variables are graphed against each other in the x-y plane. The current state of the system is represented by a point $(x(t), y(t))$ that moves around in the plane as time progresses, tracing out an *orbit*. You can see an example of orbits in a state space graph in Figure 9.1. The arrows indicate the movement of the point $(x(t), y(t))$ in time. Calculus provides us with many powerful tools for studying systems whose state changes more or less continuously throughout time. The main tool, of course, is the derivative dx/dt, which is the instantaneous rate of change of $x(t)$ with respect to time t.

A dynamical system is at *equilibrium* if it does not change over time. The time series graph for such a system is a horizontal line $x(t) = x_e$. In state space, a system at equilibrium is represented by an orbit that consists of a single stationary point (x_e, y_e) (for a two-dimensional state space). Mathematically speaking, equilibria are constant solutions of the model. Consider, for example, the *logistic population model*, which we will derive in a later chapter:

$$\frac{dN}{dt} = rN\left(1 - \frac{N}{K}\right).$$

Here, $N(t)$ is the population size at time t, and $r, K > 0$ are constant parameters. A solution $N(t)$ of the logistic model is a constant solution if and only if it satisfies

$$\frac{dN}{dt} = 0$$

for all time t. Thus, the equilibria are the roots of the *equilibrium equation*

$$0 = rN_e\left(1 - \frac{N_e}{K}\right);$$

that is, the equilibria are

$$N_e = 0, K.$$

For a discrete-time example of equilibria, consider the plant population model (1.3). A solution of this difference equation is a constant solution N_e if and only if it satisfies

$$N_{t+1} = N_t = N_e$$

for all time t. Thus, the equilibra are the roots of the equilibrium equation (or *fixed point equation*)

$$N_e = pbN_e.$$

If $bp \neq 1$, the only such constant N_e is the extinction equilibrium $N_e = 0$. If $bp = 1$, then all constant values of N_e are equilibria.

Another important notion is that of *stability*. Consider a planar pendulum. There are two equilibrium states: angle 0 (straight down) with zero velocity and angle π (straight up) with zero velocity. If the pendulum is at rest in the "down" position and is perturbed slightly away from equilibrium, it will remain close to equilibrium. The "down" equilibrium is called *stable*, because if the system starts close to it, the system stays close to it. In fact, the system asymptotically returns to equilibrium in this case, so the equilibrium is called *asymptotically stable*. If, on the other hand, the pendulum is at rest in the "up" position and is perturbed, the pendulum will move away from the "up" equilibrium. This equilibrium is called *unstable*.

Consider now a large hemispherical bowl, sitting upright on a table, with a tennis ball inside it. If the ball is placed on the bottom of the bowl with zero velocity, it will remain there. This state is therefore an equilibrium ("if it starts there, it stays there"). If the tennis ball is released, perhaps with a small velocity, near the bottom of the bowl, it will remain in the vicinity of the bottom of the bowl with small velocity, and so the equilibrium state is stable ("if it starts close, it stays close"). In fact, the tennis ball actually approaches the bottom of the bowl and its velocity approaches zero, so the equilibrium state is asymptotically stable. Now suppose the bowl is turned upside down. The top point of the bowl gives rise to an unstable equilibrium for the ball, because although it is true that "if you start there, you stay there," it is not true that "if you start close, you stay close." Now remove the bowl and set the tennis ball down on the table top with zero velocity. If the table is flat, the ball remains where you placed it. Therefore, it is at equilibrium. If you move the ball a small distance away and release it with a small velocity, still on the table top, it probably will not return to the old position, but it will not stray far. Thus, it is true that "if you start close, you stay close" to the original position; hence, the equilibrium is stable. However, since the ball does not necessarily return to the original state, the stable equilibrium is not asymptotically stable. We say it is *neutrally stable*. In fact, every point on the table top is a neutrally stable equilibrium.

Finally, we list a few phrases in English that you should always be able to translate directly into mathematics. We say y *is proportional to x* if and only if we can write

$$y = ax$$

for some constant a. We say y *is inversely proportional to x* if and only if we can write

$$y = \frac{a}{x}$$

for some constant a. *Population growth rate* is the change in population size with respect to time. In continuous time, this is the derivative dN/dt. A population's *per capita growth rate* is defined to be

$$\frac{1}{N}\frac{dN}{dt}.$$

For example, suppose a population's growth rate is proportional to the square root of the population size N, and inversely proportional to the temperature T. A mathematical model that describes this system is

$$\frac{dN}{dt} = a\frac{\sqrt{N}}{T}.$$

For another example, suppose a population's per capita growth rate is inversely proportional to the square of the population size. A mathematical model that describes this system is

$$\frac{1}{N}\frac{dN}{dt} = \frac{a}{N^2},$$

that is,

$$\frac{dN}{dt} = \frac{a}{N}.$$

1.5   STATISTICS

Deterministic models are approximations of real systems; a good model captures the *signal* (main deterministic trend) in the data. Nevertheless, the data likely will deviate somewhat from the model prediction. This deviation from the signal is called *noise* or *stochasticity*. The two main types of noise in biological data are *process error* and *measurement error* (*observational error*). The process error occurs because the real system is more complicated than the mathematical model. The measurement error occurs because the real system cannot be measured exactly. Stochasticity in ecological data can be handled with statistical methods, several of which will be addressed in this book.

There are two main types of process error in ecology: *environmental stochasticity* and *demographic stochasticity*. Stochastic events in a population can be likened to the toss of a fair coin. Imagine that a single coin is tossed for a population of animals. The outcome of the toss, although random, is the same for each individual member of the population. This is environmental stochasticity. Such extrinsic events as weather cause this type of noise. Now imagine that each animal in the population tosses its own coin. This time there is a random outcome for each individual. This is demographic stochasticity. Individual variability in intrinsic parameters such as birth and death rates cause this type of noise.

Systems in classical physics may have relatively little stochasticity, and their mathematical models can be so precise that some people call them "laws." Some social science systems, on the other hand, may have a lot of stochasticity—so much so that the signal may be swamped out by noise and mathematical modeling may be impossible. In ecology, deterministic and stochastic forces are more or less equally important. Therefore, noise should—ideally—be incorporated explicitly into a deterministic model to produce a stochastic version of the model. The interaction of deterministic and stochastic forces can give rise to a rich class of emergent dynamic phenomena that cannot occur in purely deterministic or purely random systems.

Stochasticity is modeled mathematically through the notions of *random variable* and *distribution*. Suppose you count the number of seabirds on a beach. If the number of birds is large, repeated counts of the exact same group of birds will probably yield different results due to observational error. We can use the notion of a *random variable X* to stand for the outcomes of trial observations (counts). Note that random variables are usually denoted with uppercase letters. A particular observation $X = x$, where x is an observed value, is called a *realization* of the random variable. You would want to know if some observations x are more likely than others, because you might want to know, for example, whether the count errors are biased. Mathematically, you would be asking, "what is the distribution of the random variable X?" The answer to this question is, of course, situation dependent. One has to make assumptions about how measurements, and hence errors, are distributed. Ideally, such assumptions can be tested experimentally.

The main distribution used in this book is the *normal distribution*. We will continue our discussion by supposing the bird count random variable X is normally distributed about the true number of birds μ. The normal distribution (Figure 1.2) *probability density function*

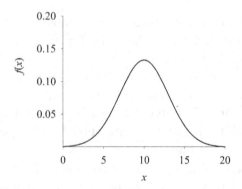

Figure 1.2: The normal distribution with $\mu = 10$ and $\sigma = 3$.

(PDF) is given by

$$f(x) = \frac{1}{\sigma\sqrt{2\pi}} e^{-\frac{1}{2}\left(\frac{x-\mu}{\sigma}\right)^2}. \qquad (1.4)$$

Here, $f(x)$ is *not* the probability of observing x birds. Rather, the probability that the observational count X will fall between the values a and b is the area under the normal curve that lies between the vertical lines $x = a$ and $x = b$; that is,

$$P\left[a \leq X \leq b\right] = \int_a^b f(x)\,dx.$$

The total area under the normal curve gives the probability that the count lies between $-\infty$ and $+\infty$, which is, of course, equal to one. Therefore, it must be true that (Exercise 10)

$$P\left[-\infty < X < +\infty\right] = \int_{-\infty}^{\infty} f(x)\,dx = 1. \qquad (1.5)$$

The *expected value* of the random variable X, often denoted $E(X)$, is the sum of all possible outcomes of X weighted by the probability of obtaining that outcome. For the normal distribution, $E[X] = \mu$; that is (Exercise 11),

$$E[X] = \int_{-\infty}^{\infty} xf(x)\,dx = \mu. \qquad (1.6)$$

The inflection points of the normal PDF f occur at $x = \mu \pm \sigma$ (Exercise 12). The parameter σ is called the *standard deviation*. It measures the spread of the distribution about the *mean* μ. The square of the standard deviation, σ^2, is called the *variance* of the distribution.

You may have noticed that we have assumed X to be a *continuous random variable*, while in the bird counting example, it is actually a *discrete random variable*. That is, you can count only an integer number of birds, while the normal distribution allows x to be any real number. We could have used a bell-shaped histogram for the distribution of a discrete random variable, but we often simplify problems with continuous approximations.

1.6   EPISTEMOLOGY: HOW WE KNOW

There are two main kinds of logical inference: *deduction* and *induction*. Deduction infers a particular conclusion from a general statement. For example,

All adult male northern cardinals (*Cardinalis cardinalis*) are red.

The bird Jim is observing is an adult male northern cardinal.

Therefore, the bird Jim is observing is red.

Notice that if the first two statements are true, then we all agree that the conclusion is guaranteed to be true. Deductive arguments are *conclusive*. However, if one (or both) of the first two statements is false, then the conclusion is not guaranteed to be true; it might be true or it might be false.

Induction infers a general conclusion from a set of particular statements or observations:

All the adult male northern cardinals I have observed are red.

Therefore, all adult male northern cardinals are red.

Notice that, despite your observations, the conclusion might still be false, depending on the sample of data you observed. Perhaps you did not have the chance to observe an albino cardinal. The only way you can be completely sure of your conclusion is if your induction was *exhaustive*, that is, if you observed *all* male northern cardinals. Inductive arguments are not, in general, conclusive unless you are able to observe all possible instances of the data.

Here, we pause to speak to fellow aficionados of Sherlock Holmes. Despite the frequent use of the word "deductive" in the Sherlockian canon, Holmes' methods were mostly inductive! We wonder how many students have missed questions about deduction versus induction on standardized exams because they thought back to the methods of that master of *induction*!

Pure mathematics is deduction. Make no mistake: The activity of doing mathematics is not purely deductive. Trial and error, hunches, and experimentation lead mathematicians to pose the conjectures that they then try to prove deductively as theorems. The final result, however, must be purely deductive, or it is not pure mathematics. Inductive arguments are not permitted as mathematical proof. (In your mathematics classes, you may have learned a special kind of proof called "mathematical induction" (MI). MI is actually a special type of deduction. See Exercise 14.) Mathematics is the only discipline in which arguments are conclusive. In one sense, deduction cannot create "new" knowledge about physical reality, since the conclusions must be implicit in the statements with which you begin. In another sense, deduction *can* create new knowledge, because it is a powerful tool that teases information out of more general statements—information that you might never have guessed was implied by the original statements. The "problem," of course, is that the general statements themselves must have come from even more general statements, and so on. You can see that in pure deduction, there must be original statements which must be taken without proof. Indeed, the edifices of pure mathematics ideally rest on unproven original statements called *axioms*. Now, pure mathematicians do not ask whether those axioms are true with respect to the observed world. They only want to know that the axioms are *logically independent* and *logically consistent*, that is, that no axiom is derivable from the others, and that they cannot generate a logical contradiction among themselves. The axiomatic method is a fascinating and tremendously powerful tool.

In general, scientists *do* want to know whether or not their starting statements correspond to observations in the real world. Unlike pure mathematicians, they are interested in the correspondence between logical inferences and observations of nature. Science, therefore, utilizes both deduction and induction. Science inductively infers hypotheses

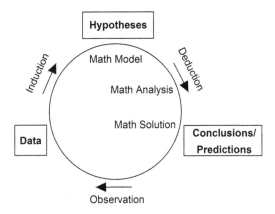

Figure 1.3: Scientific methodology.

from data, draws deductive conclusions (called predictions) from these hypotheses, and then tests the predictions against more data (Figure 1.3). The inclusion of observation and induction in scientific methodology makes science more powerful than pure mathematics alone as a tool for understanding real systems. But there is a price to pay for this increase in power, and that is a decrease in certainty. *In science, there is no proof in the mathematical sense of the word.*

The most powerful science is not purely observational or inductive, but also uses deduction. Observation guides theory, and theory guides observation. Note that the so-called *hard sciences* (physics is the prime example) employ mathematics in the deductive step (Figure 1.3).

The richness of the humanities is a hallmark of what it means to be human. The humanities ask some of the most important questions about reality, questions science and mathematics cannot address, such as questions about meaning, values, ethics, and God. The humanities use deduction and induction, but they also employ other ways of knowing. A great novel, for example, although not literally true, can be more true than any transcript of reality. A great landscape painting can convey more truth about nature than a snapshot or a data sheet. This increase in scope of questions the humanities can address comes at the price of a further decrease in certainty, in the sense of being able to convince someone else of what you know.

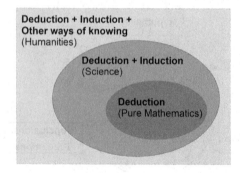

Figure 1.4: Nested epistemologies for pure mathematics, science, and the humanities.

When we compare the types of *epistemologies* (ways of knowing) of various disciplines, we can see that they are nested (Figure 1.4), as in a Venn diagram.

We also see that the more rigorous the method of inference, the more certain one can be, but the less one can learn about nature (Figure 1.5). The least rigorous methods of "inference" include other,

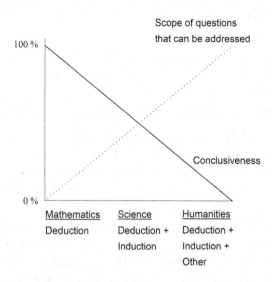

Figure 1.5: Scope and conclusiveness of the epistemologies of the disciplines.

extra-rational (not *irr*ational) ways of knowing. These epistemologies are the most "powerful," in the sense that they can address the greatest range of questions about reality. But they are also the least "compelling," in the sense of being able to rationally convince someone else. (See Exercise 26.)

In conclusion, systems of thought—mathematics, science, and the humanities—have different epistemologies. One cannot do science (using induction) and call it pure mathematics, even though it may be perfectly good science. One cannot do theology (using scripture) and call it science, even though it may be perfectly good theology. Nevertheless, all of these disciplines may have a common goal: understanding reality through discipline-specific types of models.

Science is not so much about *certainty* as it is about *understanding*. In particular, mathematical modeling is not about absolute certainty. It is about understanding. In the words of Alfred J. Lotka, an early pioneer in mathematical ecology, "An empirical formula is therefore not so much the solution of a problem as the challenge to such a solution. It is a point of interrogation, an animated question mark" (Lotka 1925).

We hope you will find this book to be an "animated question mark."

1.7 EXERCISES

1. Consider the population model

$$\frac{dx}{dt} = (b - d)\, x.$$

Identify the independent variable(s), the state variable(s), and the parameter(s).

2. Consider the LPA model of flour beetle (*Tribolium*) dynamics

$$
\begin{aligned}
L(t+1) &= bA(t)e^{-c_{el}L(t)-c_{ea}A(t)} \\
P(t+1) &= (1 - \mu_l)\, L(t) \\
A(t+1) &= P(t)e^{-c_{pa}A(t)} + (1 - \mu_a)\, A(t).
\end{aligned}
$$

Identify the independent variable(s), the state variable(s), and the parameter(s).

3. Consider the logistic population model

$$\frac{dN}{dt} = rN\left(1 - \frac{N}{K}\right).$$

Identify the independent variable(s), the state variable(s), and the parameter(s).

4. Consider the dynamical system

$$\frac{dx}{dt} = 2x(1 - x)(2x - 4)^2.$$

Find all the equilibria.

5. Consider the discrete-time dynamical system known as the Ricker model:

$$x_{t+1} = bx_t e^{-cx_t}.$$

Here, $b, c > 0$. This famous model was first used in fisheries (Ricker 1954). We will see it again several times in this book. Find all the equilibria.

6. Consider the model

$$\frac{dx}{dt} = x + xy$$
$$\frac{dy}{dt} = x + y.$$

Find all the equilibria. Hint: The equilibria are constant solution pairs (x_e, y_e) and are obtained by setting both derivatives equal to zero and solving the resulting algebraic system of equations.

7. Malthusian, or exponential, growth is modeled by

$$\frac{dx}{dt} = rx.$$

Is the extinction equilibrium $x_e = 0$ asymptotically stable, neutrally stable, or unstable? Hint: Consider the cases $r < 0$, $r = 0$, and $r > 0$ separately.

8. The per capita growth rate of a certain population is proportional to both the population size and the temperature, and it is inversely proportional to the humidity. Assume the temperature and humidity are constants on the time scale of interest. Choose variable names for the dependent and independent variables and for the parameters. Translate the statement into a mathematical model.

9. Identify the following random variables as environmental or demographic: wind speed, per capita birth rate, outdoor temperature, probability of death due to predatory encounter.

10. *Note to instructor: This exercise requires a knowledge of vector calculus (third-semester calculus).* Verify equation (1.5). Hint: Show that

$$\left(\int_{-\infty}^{\infty} \frac{1}{\sigma\sqrt{2\pi}} e^{-\frac{1}{2}\left(\frac{x-\mu}{\sigma}\right)^2} dx \right)^2$$

$$= \frac{1}{2\pi\sigma^2} \int_{-\infty}^{\infty} e^{-\frac{1}{2}\left(\frac{x-\mu}{\sigma}\right)^2} dx \int_{-\infty}^{\infty} e^{-\frac{1}{2}\left(\frac{y-\mu}{\sigma}\right)^2} dy$$

$$= \frac{1}{2\pi} \int_{-\infty}^{\infty} e^{-\frac{1}{2}u^2} du \int_{-\infty}^{\infty} e^{-\frac{1}{2}v^2} dv$$

$$= \frac{1}{2\pi} \int_{-\infty}^{\infty} \int_{-\infty}^{\infty} e^{-\frac{1}{2}[u^2+v^2]} du\, dv,$$

where $u = (x - \mu)/\sigma$ and $v = (y - \mu)/\sigma$. Now change from rectangular to polar coordinates and carry out the integration.

11. *Note to instructor: This exercise requires a knowledge of improper integrals (second-semester calculus).* Verify equation (1.6) for $\mu = 0$. That is, verify that

$$\frac{1}{\sigma\sqrt{2\pi}} \int_{-\infty}^{\infty} xe^{-\frac{x^2}{2\sigma^2}} dx = 0.$$

12. Prove that the inflection points of the normal curve (1.4) occur at $x = \mu \pm \sigma$.

13. Suppose you want to count a very large group of objects. Assume the number of objects is fixed in time. There are so many objects

that your repeated counts give different tallies. Suppose you record a large number of your repeated counts and find they are normally distributed with standard deviation s. Would you expect s to be dependent or independent of group size? Explain.

14. Mathematical induction (MI) is a particular kind of mathematical argument used to prove theorems of the form, "For all natural numbers $n = 1, 2, 3, \ldots$, the statement $Q(n)$ is true." MI is like climbing an infinitely tall ladder, where each rung is a natural number. Provided one can always move to the next highest rung from any given rung, and provided one can get on the first rung to begin with, one can ascend the whole ladder. MI uses a three-step method. First, one proves that the statement is true for $n = 1$, that is, $Q(1)$. Second, one assumes the statement is true for an arbitrary number $n = k$, that is, $Q(k)$. Third, one shows that the statement is therefore true for $n = k + 1$, that is, that $Q(k + 1)$. Here is an example: We will use MI to prove the statement "The sum of the first n natural numbers is $1 + 2 + 3 + \ldots + n = n(n+1)/2$. Proof: (i) Clearly, this formula is true for $n = 1$. (ii) Assume it is true for $n = k$; that is, assume $1 + 2 + 3 + \cdots + k = k(k + 1)/2$. (iii) Then, using part (ii), we have $1 + 2 + 3 + \cdots + k + (k + 1) = k(k + 1)/2 + (k + 1) = (k + 1)(k/2 + 1) = (k + 1)([k + 1] + 1)/2$, so the formula is true for $n = k + 1$. QED

a. Use MI to prove that the sum of the first n squares is $1^2 + 2^2 + 3^2 + \cdots + n^2 = n(n + 1)(2n + 1)/6$.

b. Is MI induction or deduction, or could it be both? Explain thoroughly.

15. Deduction is conclusive. Induction generally is not. Therefore, scientific reasoning (induction + deduction) is not conclusive, yet it is still a compelling and powerful type of argument. The following questions are highly nontrivial.

a. Why do you believe good scientific reasoning is compelling?

b. Why do you believe deduction is conclusive?

c. Why is the word "believe" used in (a) and (b)?

16. Can one use deduction to prove conclusively that deduction is conclusive?

17. Suppose you observe $x = 4,269$ birds on the beach, whereas the model predicted $x = 0$. Clearly, something is wrong, either with the model or with your observation. You conclude there is something wrong with the model. Why did your observation take precedence over the model as the standard for reality in this case? Can you prove that your observation depicts physical reality more accurately than the model prediction? Explain.

18. Criticize or support the following statement: Every system of thought must begin with unprovable presuppositions; however, a good system of thought minimizes its presuppositions.

19. Criticize or support the following statement: If two models describe the same phenomenon equally well, the most elegant model is the one closest to the truth.

20. How does the statement in Exercise 19 relate to Occam's Razor?

21. What presuppositions do all scientists have in common?

22. Criticize or support the following argument: Every system of thought begins with unprovable presuppositions; thus, no system of thought can yield absolute certainty about "reality." Therefore, all systems of thought are equally valid. Hint: Are astronomy and astrology equally valid?

23. Do you think there is a difference between *extra-rational* and *irrational*? Explain.

24. Consider Figure 1.4. Can science logically contradict mathematics? Can philosophy or theology logically contradict science (or mathematics)? Note that it can be very difficult to prove that an apparent contradiction is really a logical contradiction. Some apparent contradictions are actually *paradoxes* that can be logically resolved with more information.

25. Consider Figure 1.4. Can the epistemologies of the outer levels ever be utilized legitimately at the inner levels? For example,

can a pure mathematician use scientific induction when doing mathematics? Can a mathematician or scientist use religious scripture when doing mathematics or science? Explain your answer carefully.

26. Can one hold extra-rational knowledge with internal certainty even though the information cannot be transmitted to anyone else using rational inference? Explain. If you say "yes," give examples.

27. Criticize or support the following statement: My religion's scriptures are my set of axioms, and they trump all observations made via the senses.

BIBLIOGRAPHY

Levin, S. A. 1992. The problem of pattern and scale in ecology. *Ecology* 73:1943–1967. DOI: 10.2307/1941447. [Classic paper on scale in ecology, presented as the Robert H. MacArthur Award Lecture in 1989 in Toronto, Canada. Every student of ecology and mathematical biology should read this paper.]

Lotka, A. J. 1925. *Elements of Physical Biology*. Williams & Wilkins Company, Baltimore. [Lotka was one of the first scientists to apply mathematical methods to problems of population growth. He used mathematics from the field of physical chemistry to create "physical biology." Students interested in mathematical biology, ecology, or the history of ecology should be aware of this foundational book. In Chapter 9 you will study the famous Lotka-Volterra models of predator/prey, competition, and cooperation dynamics.]

Ricker, W. E. 1954. Stock and recruitment. *Journal of the Fisheries Research Board of Canada* 11:559–623. DOI: 10.1139/f54-039. [Classic paper introducing the Ricker model.]

CHAPTER 2

Avian Bone Growth: A Case Study

2.1 WHAT YOU SHOULD KNOW ABOUT THIS CHAPTER

In this chapter, we use bone growth data collected on a seabird colony to illustrate the major aspects of modeling methodology. The chapter is based on a paper published in the *Journal of Morphology* in 2009 by the Seabird Ecology Team, a National Science Foundation-funded research group co-directed by the authors of this textbook. This paper was co-authored by an undergraduate student (Hayward et al. 2009).

This is a crucial chapter for the student; many important ideas and techniques are introduced within the context of this case study. You should take time to master this material. Most of these techniques require computer implementation. You should complete the self-guided programming tutorial in Appendix B before beginning the exercises in this chapter.

If you feel uncomfortable with basic matrix operations (addition, subtraction, multiplication, scalar multiplication, and determinants) or need a refresher, you should read Appendix A before working through Appendix B.

Appendix C contains a concise summary of the methods of this chapter along with samples of computer code, which students may consult before attempting the exercises in this chapter.

DOI: 10.1201/9781003265382-3

2.2 SCIENTIFIC PROBLEM

Growth is a fundamental process of life. Growth of multicellular organisms involves the multiplication and differentiation of cells. Cells can be arranged in different ways to create different body shapes, just like bricks can be arranged in different ways to create different buildings.

Size constrains an organism's life. You would not be surprised to see a spider walking up a vertical wall, but you would be surprised to see a human doing this. Spider locomotion is constrained by electrostatic forces, whereas human locomotion is constrained by gravitational forces. Size makes the difference.

Shape also plays a defining role in life. A giraffe feeds on leaves high in trees. By contrast, a hippo, shaped much differently, eats different things and would find it impossible to feed like the giraffe. As anatomists say, "structure follows function."

Through *development*, multicellular organisms grow from single cells into species-typical sizes and shapes. Organs within organisms do the same thing. As organs grow, they also change shape. And as they change shape, they change (or gain or lose) function.

Bones are organs of internal support in vertebrates. As a vertebrate grows, the size and shape of its bones change as well. Different bones change in different ways. By comparing the growth of various bones, we can learn about where the animal is funneling its energy and what is happening in its life.

Glaucous-winged gulls (*Larus glaucescens*) breed in large colonies in North America's Pacific Northwest. For many years, we have studied these birds at Protection Island National Wildlife Refuge, Strait of Juan de Fuca, Washington, the USA. We noticed that newly hatched young behave differently than older juveniles, and that older juveniles behave differently than adults. So we asked a simple question: How is the development of a particular behavior related to the development of a particular bone?

2.2.1 Data

Gull chicks hatch in late June and early July on Protection Island. Many chicks die from predation, overheating, and dehydration, or are killed by neighboring adults. If they are lucky enough to survive, they will fledge at about 44 days old and reach maturity at four years old.

During the hatching period, we banded 373 newly hatched chicks (Hayward et al. 2009). Over the next several weeks, we collected 80 banded chicks that had died. Because we knew when each chick had been banded, we could tell how old it was when it died. The dead chicks ranged from 0 to 42 days old. We also collected 13 dead adults on the colony.

We shipped the dead chicks and adults back to Michigan where we prepared their skeletons—a tedious, time-consuming, and smelly task. We decided to focus on three wing bones (humerus, ulna, and carpometacarpus) and three leg bones (femur, tibiotarsus, and tarsometatarsus), because these bones facilitate the easily observed behaviors of flying and walking.

Using calipers, we measured the diaphyseal length and midshaft diameter of each of the six bones from the right side of each skeleton and recorded those data along with the chick age. Data for the humerus and ulna lengths are given in Data Set 2.1. We graphed the lengths and diameters of each of the six bones against age and noted that the bones exhibited different growth patterns.

An individual female gull typically lays an egg every other day until she has produced three eggs. The first two eggs, called the A and B eggs, are generally bigger, are the first to hatch, and produce larger, healthier chicks than the C egg. Our sample consisted of 52 (65% of the sample) C chicks, probably because they are least likely to survive. Thus, our data set is skewed in favor of smaller birds, something that should be kept in mind when interpreting the results.

It is also important to note that we used bones from different birds for each age level—in other words, ours was a *cross-sectional* study. Had we followed the growth of individual bones in individual birds, we would have done a *longitudinal* study. Cross-sectional studies are limited by the fact that they tell us nothing about individual rates of growth, but only provide estimates of mean rates of growth for a population. Although it would be more informative, carrying out a longitudinal study on bone growth in wild gulls would be logistically difficult.

2.3 TRANSLATION INTO MATHEMATICS

How do gull bones grow? A bit of thought will convince you that this question cannot be translated into the language of mathematics, because it is too vague. What does "grow" mean? We might define an object to be "growing" if and only if its size is changing over time. (This definition for growth includes shrinking as well as expanding.) But what does "size" mean? Are we interested in length, diameter, volume, or what? And which kind of bone are we talking about? Humerus? Ulna?

You can see that the very first step in the modeling process, which is the translation into mathematics, typically requires careful thought. This step can be quite fruitful in and of itself, even if you never go any further in the modeling cycle, because the act of translating forces you to clarify concepts and sharpen questions. Translation into mathematics can help you ask whether your scientific question makes sense and whether it can be expected to have a solution. This is important, because some apparently meaningful questions are actually nonsensical (Exercise 3). Indeed, some of the burning scientific questions of history have simply "gone away" because they were discovered to be meaningless.

Let's pose our problem precisely. How does the length of the humerus change in time over the life of a gull? That is, how does the length of the humerus change as a function of age? Consider a single "average" gull. Let

$$
\begin{aligned}
x &= \text{Age in days} \\
f(x) &= \text{Length of humerus in cm.}
\end{aligned}
$$

Mathematically, the question becomes: How does $f(x)$ depend on x?

2.3.1 Simplifying Assumptions

When we relate the variables x and $f(x)$ through an equation, we are making implicit biological assumptions. As far as possible, these assumptions should be stated explicitly in the language of biology. Ideally, the assumptions should address two kinds of questions. First, what deterministic mechanisms are most important in driving the system? Second, what kind of process error is in the system, and what kind of measurement error is in the data?

2.3.1.1 Deterministic Assumptions

We pose two alternative hypotheses as the deterministic modeling assumption. They will give rise to two competing deterministic models, which we will test against one another using the data.

(A1a) The length of the humerus increases linearly with age until the chick reaches some critical age, at which point the bone abruptly ceases to grow.

(A1b) The length of the humerus grows first at an increasing rate and then at a decreasing rate, asymptotically leveling off (*saturating*) toward some maximal length. (This is called *sigmoidal growth* because the length vs. age curve is s-shaped.)

2.3.1.2 Stochastic Assumptions

We also make assumptions about process and measurement errors in the system:

(A2) The dominant type of noise in the system is demographic stochasticity; environmental stochasticity and measurement error are negligible.

2.3.2 The Deterministic Model

Assumption (A1a) says that humerus length $f(x)$, when graphed against age x, is a line with positive slope from age 0 (hatching) to some critical age, at which point the graph of $f(x)$ becomes a horizontal line. Such a graph can be completely determined by three parameters. We will certainly want parameters for the critical age and the maximal bone length. The third parameter could be either the y-intercept (humerus length at hatching) or the slope (rate of growth). Let

$$
\begin{aligned}
b &= \text{Age at which humerus growth stops} \\
K &= \text{Maximal humerus length} \\
a &= \text{Slope (rate of growth).}
\end{aligned}
$$

Then assumption (A1a) can be translated into mathematics as the model (Exercise 4)

$$f(x) = \begin{cases} a\,(x-b) + K & x < b \\ K & x \geq b. \end{cases} \tag{2.1}$$

Here, x is the age in days, $f(x)$ is the length of the humerus in cm at age x, and $a, b, K > 0$ are parameters.

The alternative assumption (A1b) is not specific enough to yield a unique equation, because there are several classes of functions that produce sigmoidal curves. In this example, we will assume a *Janoschek curve* (Gille and Salomon 1995)

$$f(x) = K - (K - a)e^{-bx^c}, \tag{2.2}$$

where $a, b, c, K > 0$ are parameters.

The two deterministic models (2.1) and (2.2) are competing hypotheses that we are setting forth to explain the humerus data in Data Set 2.1. It is important to note that the parameters a and b have different biological meanings in the two different models, whereas the parameter K has the same interpretation in both models. Note also that model (2.1) has three parameters, whereas model (2.2) has four parameters. The number of parameters will be important during model selection.

2.3.3 The Stochastic Model

We wish to explicitly model the stochasticity in the system based on assumption (A2) about the source of the noise. Let $F(x)$ be a random variable denoting the measurement of the humerus length in a chick of known age x. We can think of the random variable $F(x)$ as the deterministic prediction $f(x)$ plus a random perturbation (noise):

$$F(x) = f(x) + \text{noise}. \tag{2.3}$$

Think of equation (2.3) as the *deterministic skeleton* $f(x)$ "clothed" with noise. The deterministic skeleton of a stochastic model is the part of the model that would remain if all the noise could be tuned to zero.

Typically, however, noise is not additive as in equation (2.3). Usually, one must first transform the observational data and the deterministic predictions with a *variance-stabilizing transformation* ϕ under which noise becomes additive:

$$\phi\left(F\left(x\right)\right) = \phi\left(f\left(x\right)\right) + \text{noise}. \tag{2.4}$$

Here, we mention an important point from statistical theory: *Demographic noise is approximately additive on the square root scale, whereas environmental noise is approximately additive on the log scale* (Cushing et al. 2003). That is, if demographic noise is dominant, then $\phi(\cdot) = \sqrt{\cdot}$, and if environmental noise is dominant, then $\phi(\cdot) = \ln(\cdot)$.

Thus, in our current example, under assumption (A2), equation (2.4) becomes

$$\sqrt{F\left(x\right)} = \sqrt{f\left(x\right)} + \sigma\varepsilon, \tag{2.5}$$

where $\sigma > 0$ is a parameter representing the standard deviation of the noise and ε is a *standard normal random variable* (a normal random variable with mean zero and standard deviation one). Equation (2.5) is the *stochastic model* for our current example.

Let the pair of numbers (x, l_x) be an actual data point, that is, a measurement l_x of the length of the humerus in a particular chick of age x. Since we are thinking of equation (2.5) as a surrogate for the real system, we can think of l_x as a realization of the random variable $F(x)$. The *residual error* of the deterministic prediction is the actual measurement minus the predicted value. The residuals corresponding to equation (2.3) are the values of

$$res = l_x - f(x).$$

The residuals corresponding to the more general equation (2.4) are the values of

$$res = \phi\left(l_x\right) - \phi\left(f(x)\right),$$

and the residuals corresponding to our particular example, with equation (2.5), are the values

$$res = \sqrt{l_x} - \sqrt{f(x)}.$$

Note that any particular residual is a realization of the random variable $\sigma\varepsilon$ in the stochastic model (2.5). Thus, *on the square root scale the residuals are hypothesized to be normally distributed with mean zero and standard deviation σ.*

2.4   MODEL PARAMETERIZATION

It might be difficult to estimate the maximal adult humerus length K from chick data, because none or very few of the dead chicks might have approached maximal humerus length. From the 13 dead adults, however, we have explicit data on the maximal humerus length K. The lengths of the 13 adult humeri are listed at the bottom of Data Set 2.1. We cannot include these adult data in estimating parameters a, b, and c, because the ages of the adult birds were unknown. However, we can use the adult data to estimate directly a value for K, thus reducing the number of model parameters by one. A statistically sophisticated way to estimate K is found in Hayward et al. (2009); here, however, we simply estimate the value of K as the sample mean of the adult humerus data in Data Set 2.1:

$$\widehat{K} = 11.96923077.$$

The "hat" in $\widehat{K}$ indicates that this is the estimated value of K. We will save quite a few decimals for the computations that follow, even though they are not all significant. *In general, only the results of final computations should be rounded.*

Our next goal is to use the juvenile data to estimate a, b, and c.

2.4.1   Dividing the Data Set

Which data are to be used for parameter estimation? This often requires careful thought; the choice depends on the scientific question being asked, as well as how much and what kind of data exist. In general, if the data set is fairly large and robust, it is best to set aside (randomly) some of the data for purposes of independent model validation.

We will divide the juvenile data set in Data Set 2.1 into two parts, one for model fitting (parameterization) and one for model evaluation (validation). We will call the two data sets the *estimation data set* and the *validation data set*. We want to divide the data randomly, but at the same time, we would like to have representative data points from a variety of ages in each data set. We therefore use a technique called *stratified random sampling* in the following steps:

1. Chick ages in Data Set 2.1 range from $x = 0$ to $x = 42$ with integer values. *Bin* the juvenile humerus data into the age

intervals $[0, 4]$, $[5, 9]$, $[10, 14]$, ..., $[35, 39]$, $[40, 44]$ (Data Set 2.2). Our goal is to randomly select half of the data from each bin.

2. Number the data points in each bin (Data Set 2.2).

3. Consider the first bin. It has 18 data points numbered 1 through 18. Use a random number generator to flip an "18-sided coin" 50 times, and write down the sequence of random numbers generated. That is, use the computer to generate a list of 50 integers from 1 to 18 chosen from a uniform random distribution. Repeat this step for each bin. Such a procedure yielded the sequences of numbers in Data Set 2.3. (Obviously, if you carried this out again, you would get a different set of sequences of random numbers.)

4. Consider the random sequence for the first bin: 17, 12, 16, 2, 15, 8, 16, 13, 12, 6, 3, 3, ... Put data point number 17 into data set I, number 12 into data set II, number 16 into data set I, number 2 into data set II, number 15 into data set I, number 8 into data set II, (skip over the 2nd occurrence of number 16), number 13 into data set I, (skip over the 2nd occurrence of number 12), number 6 into data set II, etc. Skip over any random number that has already occurred in the sequence. Do this for each bin. The data are now divided into two sets (I and II as shown in Data Set 2.3). Notice that if a bin had an odd number of data points, then data set I will always have one more data point than data set II, which is not a problem.

5. Flip a (two-sided) coin for Bin 1. "Heads" means data set I for that bin goes into the estimation data; "tails" means it goes into the validation data. Do this for each bin. Such a procedure yielded the data sets in Data Set 2.4.

The data are now divided. The estimation and validation data are graphed together in Figure 2.1. Make sure you can reconstruct Data Set 2.4 from the original Data Set 2.1, given the sequences of random numbers in Data Set 2.3 (Exercise 6).

Our next goal is to use the estimation data to estimate the parameters (other than K). In what follows, we discuss two methods for doing this.

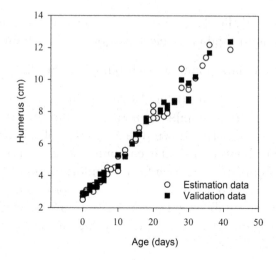

Figure 2.1: Estimation and validation data shown as length vs. age.

2.4.2 Maximum Likelihood (ML) Method

Our model assumes that the residual errors $res = \sqrt{l_x} - \sqrt{f(x)}$ are normally distributed with mean zero and standard deviation σ. Thus, the likelihood that a given data point (and hence a given residual) occurred as a realization of our stochastic model (2.5) is

$$\frac{1}{\sigma\sqrt{2\pi}}e^{-\frac{1}{2}\left(\frac{res}{\sigma}\right)^2}.$$

This is the normal distribution PDF from Chapter 1. The likelihood that the entire data set occurred as independent realizations of our stochastic model is the product of the likelihoods over the whole data set:

$$L = \prod_{\text{data}} \frac{1}{\sigma\sqrt{2\pi}}e^{-\frac{1}{2}\left(\frac{res}{\sigma}\right)^2}, \tag{2.6}$$

where the symbol Π denotes the product operator. If there are q residuals, then by the properties of exponents, equation (2.6) is equivalent to (Exercise 7)

$$L = \left(\frac{1}{\sigma\sqrt{2\pi}}\right)^q \exp\left(-\frac{1}{2\sigma^2}\sum_{\text{data}}(res)^2\right), \tag{2.7}$$

where $\exp y$ means e^y.

We want to *maximize the likelihood L that the whole data set occurs as a set of independent realizations of our stochastic model.* The only things that can be adjusted on the right-hand side of equation (2.7) are the residuals $res = \sqrt{l_x} - \sqrt{f(x)}$, which depend on the deterministic model prediction $f(x)$. The values of $f(x)$ can be adjusted only by changing the values of the model parameters. Thus, for model (2.1), the likelihood L is a function of parameters a and b:

$$L(a,b) = \left(\frac{1}{\sigma\sqrt{2\pi}}\right)^q \exp\left(-\frac{1}{2\sigma^2}\sum_{\text{data}}(res)^2\right). \qquad (2.8)$$

Equation (2.8) is called the *likelihood function.* Note that for equation (2.2), L is a function of three parameters, $L(a,b,c)$. For simplicity, we will continue to write $L(a,b)$ in the explanation that follows. The goal is to maximize $L(a,b)$ as a function of the parameters a and b. We want to find the *maximizer,* that is, the pair of parameters $\left(\widehat{a},\widehat{b}\right)$ that maximizes the function $L(a,b)$. These maximizing parameters are called the *maximum likelihood* (ML) *parameter estimates.*

Note that a maximizer for L will also be a maximizer for $\ln L$, and vice versa (Exercise 8). The properties of logarithms allow us to write the log-likelihood function as (Exercise 9):

$$\ln L(a,b) = -q\ln\sigma - \frac{q}{2}\ln(2\pi) - \frac{1}{2\sigma^2}\sum_{\text{data}}(res)^2. \qquad (2.9)$$

Thus, the ML parameter estimate $\left(\widehat{a},\widehat{b}\right)$ is the parameter vector that maximizes $\ln L(a,b)$.

2.4.3 Nonlinear Least Squares (LS) Method

Nonlinear least squares (LS) parameter estimates are obtained by minimizing the *residual sum of squares* (RSS)

$$\text{RSS}(a,b) = \sum_{\text{data}}(res)^2 \qquad (2.10)$$

as a function of the parameters (a,b). The LS parameter estimate $\left(\widehat{a},\widehat{b}\right)$ is the parameter vector that minimizes RSS.

Although it would be more accurate to say "sum of squared residuals" than to say "residual sum of squares," RSS is the traditional notation for equation (2.10).

Note that in our current example having one state variable, in which noise is Gaussian (normally distributed) with mean zero and constant variance σ^2 and the residuals are independent, maximizing the log-likelihood (2.9) is equivalent to minimizing the RSS (2.10). (See Exercise 10.)

One nice thing to know about LS, however, is that it loosens the restrictive assumptions on the residuals; LS parameter estimates converge to the true values even if the noise is non-normal and autocorrelated, as long as the noise has a stationary distribution (Cushing et al. 2003; Tong 1990).

Once a and b are estimated, the variance of the residuals is estimated by

$$\widehat{\sigma}^2 = \frac{\widehat{RSS}}{q},$$

where $\widehat{RSS}$ denotes the fitted value of RSS.

2.4.4 Downhill Minimization Routine: Nelder-Mead Algorithm

Typically, one cannot maximize the log-likelihood or minimize RSS analytically; it must be done numerically on a computer. For equation (2.1), think of RSS(a, b) as a surface suspended over *parameter space*. In this case, parameter space is the horizontal plane spanned by the a-axis and b-axis. The RSS-axis rises vertically out of the plane. Each point on the plane corresponds to a parameter pair (a, b), and the value of RSS(a, b) is plotted above each point on the plane, generating a surface. We want to locate the *minimizer* point(s) $\left(\widehat{a}, \widehat{b}\right)$ on the plane at which the surface attains a minimum value.

The Nelder-Mead algorithm begins with an initial "guess" (a_0, b_0) in parameter space and then systematically checks around nearby in parameter space to see if there is a lower point on the surface. In this way, the routine "walks downhill" along the RSS surface and converges on a local minimum. It is important to remember that a surface can have more than one local minimum. *There is no foolproof way of making sure you have found a global minimum. Always try a variety of different initial guesses to see if the Nelder-Mead algorithm always converges to the same minimum.*

2.4.5 Implementing Parameterization in Code

A code to estimate model parameters requires three basic parts: a main program, a downhill search function, and a program that computes the RSS.

The main program is a "front end" in which the user sets the initial "guess" for the parameter vector. This program passes the initial vector and the name of the RSS subroutine to the downhill search algorithm (which attempts to converge on a minimizer for RSS) and, finally, outputs the minimizer (parameter estimates) that it obtained from the downhill algorithm.

The second part is the downhill search program. This is not something you need to code yourself; it is likely in the library of functions of the language you are using. For example, in MATLAB the downhill search function is called *fminsearch*. When the main code passes the initial parameter vector and the name of the RSS subroutine to this search function, it uses those initial parameters to compute RSS via the RSS subroutine. It then chooses a suite of nearby parameters, computes RSS for them, and uses the parameters associated with the smallest RSS as the next "guess." It continues this process of "walking downhill" until it finds a local minimum value of RSS. Finally, it returns the minimizer vector of parameters to the main code.

The third part is the subroutine that computes the RSS as a function of the parameters that are passed to it by the search algorithm. This is where the model equations are coded, predictions and residuals are computed, and the RSS value is computed.

Most downhill search functions search for negative as well as positive minimizers. In population models, however, parameters are usually positive and we do not want to search negative parameter space. One way to address this is to pass the initial parameter vector v to the downhill routine as the vector $\ln v$ so that it can search all of parameter space. This works because the log function maps the interval $(0, \infty)$ to the interval $(-\infty, \infty)$. In the RSS routine, the vector $\ln v$ must be converted back to v via exponentiation before the parameters are used to compute model predictions. Also, when the final parameter estimates are returned to the main program from the search algorithm, they are returned as $\ln \widehat{v}$, and so they must be exponentiated in order to recover the parameter estimates $\widehat{v}$ on the correct scale.

TABLE 2.1 LS Parameters for Equations (2.1) and (2.2) Estimated from the Estimation Data on the Square Root Scale

Model	$\widehat{a}$	$\widehat{b}$	$\widehat{c}$
A1a	0.2475	37.92	N/A
A1b	2.872	0.005402	1.661

The best way to understand all this is to work through sample code. For codes similar to the ones you will write in the exercises concerning the current example, see Appendix C.

2.4.6 Results of Parameterization

The LS parameters for models (2.1) and (2.2), estimated from the estimation data on the square root scale under the assumption that $\widehat{K} = 11.96923077$, are shown in Table 2.1 to four significant figures (Exercise 11).

The first question we should ask is: Do these parameters make biological sense (Exercise 12)? *Note that if one or more parameters persist in wandering off to zero or infinity during the parameterization process no matter what starting parameter values you choose, then the model is probably flawed and should be revised.*

The predictions of both models at their LS parameter values are shown with the estimation data in Figure 2.2a and b. Which of the two alternative models best fits the data? Is it clear visually, or do we need a quantitative method to determine which model is best? We turn to this question in the next section.

2.5 MODEL SELECTION

Which of the two alternative models, (2.1) or (2.2), best describes the data? The two models have different numbers of parameters. A curve with many parameters is more flexible and easier to fit to data than one with few parameters. Indeed, an overparameterized model can fit just about any data, but has little explanatory power. Thus, the model with more parameters should be penalized for overfitting. To do this,

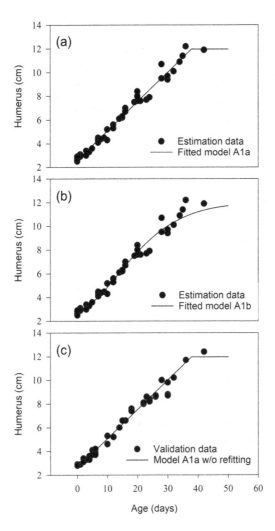

Figure 2.2: Model predictions. (a) Model (A1a) as fitted to the estimation data. (b) Model (A1b) as fitted to the estimation data. (c) Model (A1a) as compared to the validation data without refitting.

researchers often use a tool from information theory called the *Akaike information criterion* (AIC) to select the best model (Burnham and Anderson 2002). The AIC is defined as

$$\text{AIC} = -\ln\left(\widehat{L}\right) + 2\kappa, \tag{2.11}$$

where $\widehat{L}$ is the maximized value for the likelihood function and κ is the number of estimated parameters in the model. Model comparison is based on the rank of the AIC values for the suite of alternative models. The least AIC indicates the best model. The first term in equation (2.11) is smallest for the model with the largest likelihood, but this is discounted by the second term, which is largest for the model with the most parameters.

In our current example, the residuals are Gaussian with mean zero and constant variance σ^2. In such a situation, it is possible to show that ranking the AIC values (2.11) is equivalent to ranking the values of

$$\text{AIC}^* = q \ln \widehat{\text{RSS}} + 2\kappa, \tag{2.12}$$

where q is the number of residuals, $\widehat{\text{RSS}}$ is the fitted value of RSS, and κ is the number of parameters being estimated, including σ (Exercise 18).

Note that if all of the alternative models have the same number of parameters κ, then selecting the model with minimum AIC is equivalent to selecting the model with the maximum likelihood (for ML estimation) or the minimum RSS (for LS estimation), in which case there is no reason to compute the AIC.

In our current example, using equation (2.12), we obtain the AIC values shown in Table 2.2 (Exercise 16). The least AIC belongs to model (2.1). Thus, we discard model (2.2) in favor of model (2.1). Our deterministic model is therefore

$$
\begin{aligned}
f(x) &= \begin{cases} a\,(x-b) + K & x < b \\ K & x \geq b \end{cases} \\
a &= 0.2475 \\
b &= 37.92 \\
K &= 11.97,
\end{aligned}
\tag{2.13}
$$

TABLE 2.2   AIC Values Obtained from Equation (2.12)

Model	q	$\widehat{\sigma}^2$	κ	AIC*	Δ AIC
A1a	40	0.005266	3	−203.9	0.0000
A1b	40	0.006106	4	−195.9	8.000

and our demographic stochastic model is

$$\sqrt{F(x)} = \sigma\varepsilon + \begin{cases} \sqrt{a(x-b)+K} & x < b \\ \sqrt{K} & x \geq b \end{cases} \tag{2.14}$$

$$a = 0.2475$$
$$b = 37.92$$
$$K = 11.97$$
$$\sigma = \sqrt{0.005266}.$$

2.6 MODEL VALIDATION

To validate the selected model (2.13), we must evaluate how well it explains the validation data, *without reparameterizing*. We will compute a number called the *goodness of fit* on both the estimation data set and the validation data set. If the model fits the validation data as well as the estimation data, then we will consider the model validated.

One measure of goodness of fit is the generalized *R-squared* value, denoted R^2. Consider a data set with members (x, l_x). Let $\overline{l_x}$ be the sample mean of the data set. Then the R-squared value (when data are not transformed by ϕ) is defined to be

$$R^2 = 1 - \frac{\sum_{\text{data}} (l_x - f(x))^2}{\sum_{\text{data}} (l_x - \overline{l_x})^2}.$$

The residual sum of squares $\sum_{\text{data}} (l_x - f(x))^2$ is in the numerator; it measures the variability that is unexplained by the model. The denominator $\sum_{\text{data}} (l_x - \overline{l_x})^2$ measures the variability around the mean of the data. Thus, the quotient of these two numbers measures the fraction of variability not explained by the model, relative to the use of the mean as a predictor. One minus the quotient measures the fraction of variability that *is* explained by the model.

In general, for transformed data, the R-squared is

$$R^2 = 1 - \frac{\sum_{\text{data}} (\phi(l_x) - \phi(f(x)))^2}{\sum_{\text{data}} \left(\phi(l_x) - \overline{\phi(l_x)}\right)^2},$$

TABLE 2.3 R^2 Values for Model (2.13) as
Fitted to the Estimation Data and Computed
on the Validation Data without Refitting

	Estimation Data	Validation Data
R^2	0.9838	0.9813

where $\overline{\phi\left(l_x\right)}$ denotes the mean of the transformed data. For our current
example, we have

$$R^2 = 1 - \frac{\sum_{\text{data}} \left(\sqrt{l_x} - \sqrt{f(x)}\right)^2}{\sum_{\text{data}} \left(\sqrt{l_x} - \overline{\sqrt{l_x}}\right)^2},$$

where $\overline{\sqrt{l_x}}$ denotes the mean of the square roots of the humerus length
measurements.

Note that it is always true that $R^2 \leq 1$. The closer R^2 is to one,
the better the model fit. Although R^2 is usually between zero and one,
it can also be negative (Exercise 15).

The R^2 values for model (2.13) on the two data sets are given
in Table 2.3 (Exercise 17). Successful validation is supported because
the R^2 values are about the same. The predictions of model (2.13) are
shown in Figure 2.2a and c with the estimation and the validation data,
respectively. Visually, note that the model appears to fit the validation
data about as well as it fits the estimation data set upon which it was
parameterized.

2.7 EXERCISES

1. Work through the self-guided linear algebra tutorial in Appendix
   A.

2. Work through the self-guided programming tutorial in Appendix
   B.

3. Some questions that seem to make sense are actually nonsensical.
   Explain why each of the following questions is nonsensical.

   a. What is the basal metabolism rate of a jackalope?

   b. What proportion of the mass of iron is due to phlogiston?

    c. How many minutes are there in a kilometer?

    d. Let A be the set of all sets which do not have themselves as a member, that is, $A = \{B \mid B \notin B\}$. Is $A \in A$?

4. Given only the verbal description in assumption (A1a), how would you write down a mathematical model equivalent to that description? In other words, derive model (2.1) from assumption (A1a). Explain each step as you go, and explain why the model has to have at least three parameters.

5. Another well-known sigmoidal curve is the Holling type III curve, which has the form

$$y = \frac{Mx^2}{a^2 + x^2}; \quad x \in [0, \infty); \quad a, M > 0. \qquad (2.15)$$

Use calculus to do the following problems.

    a. Show that $y(0) = 0$ and $\lim_{x \to \infty} y(x) = M$.

    b. Show that $y(x)$ is an increasing function of x.

    c. Where is y concave up? Concave down? Find all inflection points.

    d. Find the value of x at which the curve reaches one half of its asymptotic limit. This is called the *half-saturation constant*.

    e. Use the information gained above to graph y vs x.

    f. The Holling type III curve cannot describe the length of the humerus, because the humerus has a positive length b upon hatching; that is, $f(0) = b$. We want to translate the Holling type III curve vertically so that the initial value is b instead of zero. Also, we want the saturation level to be K. From equation (2.15), derive the following modification of the Holling III model:

$$f(x) = \frac{(K - b) x^2}{a^2 + x^2} + b.$$

    Explain each step as you go.

6. Make sure you can reconstruct Data Set 2.4 from the original Data Set 2.1, given the sequences of random numbers in Data Set 2.3. You don't need to show any work for this problem, but indicate whether or not you were successful in reproducing Data Set 2.4.

7. Show that equation (2.6) can be written as

$$L = \left(\frac{1}{\sigma\sqrt{2\pi}}\right)^q e^{-\frac{1}{2\sigma^2}\sum_{\text{data}}(res)^2},$$

where q is the number of residuals.

8. Let $g(z)$ be any positive-valued function that is twice differentiable. Use the "derivative tests" from calculus to prove that a maximizer for $g(z)$ will also be a maximizer for $\ln g(z)$, and vice versa. Can you drop the "twice differentiable" hypothesis and still prove the result?

9. Derive the log-likelihood function (2.9) from the likelihood function (2.8).

10. Explain why the log-likelihood in equation (2.9) is maximized when the residual sum of squares (equation 2.10) is minimized.

11. Reproduce the LS parameter estimates in Table 2.1. Attach your programs, input files, and screenshots of your output.

12. Can you determine the biological meanings of the parameters in each of models (2.1) and (2.2)? Are the LS parameter estimates in Table 2.1 for the humerus model (2.13) biologically reasonable? Explain. Hint: It may take some work to thoroughly understand the biological meanings of parameters b and c in the Janoschek model (2.2). Reading the 1995 paper by Gille and Salomon (1995) will help.

13. Stochastic models are important for two reasons. First, they are necessary for parameterization because they specify how the residuals are distributed. Second, they can be used to simulate the real (noisy) system. Write a program to produce a simulated data set of humerus lengths using the stochastic model (2.14).

For each of the ages 0, 1, 2, ... , 50, generate 10 simulated data points. Present the output as a scatter plot of length against age. Attach your program and the output graph.

14. What does $R^2 = 1$ mean? What does $R^2 = 0$ mean? Can R^2 be greater than one? Explain.

15. Consider the number of seals hauled out on a beach at hour t. Suppose a wildlife refuge biologist models the number of seals $N(t)$ on the beach at hour t with the equation

$$N(t) = 3t - 7.$$

The biologist then collects 10 hours of data:

Time t	Seals	Model Prediction $N(t)$
5	10	
6	9	
7	11	
8	10	
9	10	
10	9	
11	8	
12	10	
13	11	
14	12	

a. Fill in the model predictions for each hour.

b. Compute R^2 assuming demographic noise.

c. Under what mathematical conditions is R^2 negative?

d. What does a negative R^2 mean to a scientist?

16. Reproduce all the values in Table 2.2. Attach your programs, input files, and screen shots of your output.

17. Reproduce the R^2 values in Table 2.3. Attach your programs, input files, and screen shots of your output.

18. If the residuals for each of the alternative models are independent and Gaussian with mean zero and constant variance, prove that ranking the AIC values from equation (2.11) is equivalent to ranking the values of

$$\text{AIC}^* = q \ln \widehat{\text{RSS}} + 2\kappa,$$

where q is the number of residuals, $\widehat{\text{RSS}}$ is the fitted value of RSS, and κ is the number of parameters being estimated, including σ.

19. Project: Model the growth of the ulna in glaucous-winged gulls. That is, repeat the comprehensive analysis in this chapter for the ulna length data in Data Set 2.1. Take your two alternative deterministic models to be the modified linear model

$$f(x) = \begin{cases} a\,(x - b) + K & x < b \\ K & x \geq b \end{cases}$$

and the modified Holling type III model

$$f(x) = \frac{(K - b)\,x^2}{a^2 + x^2} + b.$$

Use the same binning procedure as in Data Set 2.2, and use the same sequences of random numbers in Data Set 2.3, but work through all the details. Present your work in a complete, precise, and organized fashion.

20. Project: We might suspect that the humerus in C chicks grows more slowly than in A and B chicks. Given the work already done in this chapter, we can now test this hypothesis. The idea is to fit model (2.1) to A and B chick data only and then see if it also fits the C chick data without reparameterizing.

   a. Divide the humerus data in Data Set 2.1 into two sets: data set AB (consisting of all the A and B chick data) and data set C (consisting of all the C chick data).

   b. Estimate the LS parameters for model (2.1) on the AB data set. Record the parameter values and the RSS. Compute $\hat{\sigma}^2$ and R^2.

   c. Without re-estimating the parameters, compute the R^2 on the independent data set C.

   d. What is your conclusion? Does the humerus grow differently in C chicks?

   e. Present your work in a short scientific paper. Your paper should have standard sections including title page, abstract, introduction, methods, results, discussion, conclusion, and references, as well as tables and graphs. Include your code at the end of the paper in an appendix. Ask your instructor which journal style you should follow in the preparation of the manuscript.

BIBLIOGRAPHY

Burnham, K. P. and Anderson, D. R. 2002. *Model Selection and Multi-Model Inference: A Practical Information-Theoretic Approach*, 2nd ed. Springer, New York. [This is a practical, user-friendly book that shows scientists how to make valid inferences from empirical data within an information-theoretic framework.]

Cushing, J. M., Costantino, R. F., Dennis, B., Desharnais, R. A., and Henson, S. M. 2003. *Chaos in Ecology: Experimental Nonlinear Dynamics*. Academic Press, San Diego, CA. [In Chapter 6 we will draw heavily on this book to showcase how dynamic population data can be connected to mathematical models.]

Gille U. and Salomon F. V. 1995. Bone growth in ducks through mathematical models with special reference to the Janoschek growth curve. *Growth, Development and Aging* 59:207–214. [The authors compare five alternative models for bone (weight) growth in white Pekin ducks: Janoschek, Richards, Bertalanffy, Gompertz, and Logistic.]

Hayward, J. L., Henson, S. M., Banks, J. C., and Lyn, S. L. 2009. Mathematical modeling of appendicular bone growth in glaucous-winged gulls. *Journal of Morphology* 270:70–82. DOI: 10.1002/jmor.10669. [This is the paper on which Chapter 2 is based. Both theoretical and empirical: models connected to data.]

Tong, H. 1990. *Non-Linear Time Series: A Dynamical System Approach*. Oxford University Press, Oxford. [Valuable introduction to the analysis of nonlinear time series.]

II

Discrete-Time Models

II

Discrete-Time Models

CHAPTER 3

Discrete-Time Maps

3.1 WHAT YOU SHOULD KNOW ABOUT THIS CHAPTER

Think about almost any area of biology—physiology, genetics, development, ecology, or evolution. Could we study this area without considering time? Certainly not. Look at any biology text or journal. How much could you learn if you ignored all the time-based graphs? Not much. Life happens in time. Biologists, then, concern themselves with many time-based, or *temporal*, aspects of life.

Not surprisingly, mathematical modeling in biology often involves equations that describe and predict temporal processes. Some biological processes, like bone growth, occur continuously in time. Others, like population growth in annual plants, occur more discretely in time. Even though data collected from continuous processes are by necessity taken at discrete time intervals, model predictions based on these data assume temporal continuity and are commonly written in the form of differential equations. By contrast, discrete-time processes are modeled using *difference equations*, or *maps*, equations that take into account the discontinuous nature of these processes.

In this chapter, we focus on difference equations. Although difference equations themselves do not involve derivatives, the tools you need in order to study them are from first-semester calculus.

3.2 COMPARTMENTAL MODELS

Let x_t be the state of a system—say, the size of a population—at time t. Think of the population as a compartment having inflows and outflows.

DOI: 10.1201/9781003265382-5

Inflows are due to births and immigration. Outflows are due to deaths, emigration, harvesting, etc. (Figure 3.1).

In general, *compartmental models* state that the net rate of change of the quantity in a compartment is equal to the sum of its inflow rates minus the sum of its outflow rates:

$$\text{Net rate of change} = \sum \text{Inflow rates} - \sum \text{Outflow rates}.$$

For continuous-time processes, the net rate of change of a quantity $x(t)$ is the derivative dx/dt. What is the analogous concept for a discrete-time process? In mathematics, the uppercase Greek letter delta (Δ) usually stands for "change." We define Δx_{t+1} to be the amount x changes from time t to time $t + 1$. That is,

$$\Delta x_{t+1} = x_{t+1} - x_t.$$

Now, in general, Δx_{t+1} is not a rate of change; it is simply an amount of change. However, in the context of discrete time steps, Δx_{t+1} is the rate of change *per unit time step*.

Thus, a discrete-time compartmental model has the form

$$\Delta x_{t+1} = \sum (\text{Inflow rates at time } t) - \sum (\text{Outflow rates at time } t),$$
$$(3.1)$$

where each flow rate on the right-hand side of equation (3.1) is replaced by a mathematical expression based on the modeling assumptions.

In discrete-time modeling, the time step is chosen to be appropriate for the application. Discrete-time models can be particularly useful if there is a temporal "pulse" in the biological system, in which case you would choose the time step to equal the time between pulses. For example, suppose you were studying an annual plant species that

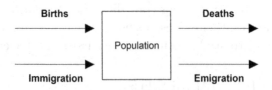

Figure 3.1: Flows into and out of a population may be due to births, immigration, deaths, and/or emigration.

produced seeds each autumn, and you wanted to track the population size. The model time step might be one year, measured in the autumn after the seeds were produced in order to take advantage of the natural reproductive pulse in the system.

3.3 LINEAR MAPS

3.3.1 Malthusian Growth

We illustrate linear maps with the following problem. Let x_t be the number of individuals in a population at time t. Assume

(A1) The average *per capita* fecundity (birth rate) is constant and positive.

(A2) Generations are nonoverlapping. All deaths occur at the end of the generational time period.

(A3) The only flows into or out of the population are due to births and deaths.

We need a parameter for the per capita birth rate. Let

b = Number of births per individual per unit time.

Since the per capita birth rate is b, the total population birth rate will be b times the number of individuals in the population:

bx_t = Number of births in the population per unit time at time t.

The population death rate is exactly x_t individuals per unit time because all individuals die at the end of each time step. Following equation (3.1), we can write

$$\Delta x_{t+1} = \text{Birth rate at time } t - \text{Death rate at time } t \quad (3.2)$$
$$= bx_t - x_t.$$

By definition of Δx_{t+1}, equation (3.2) becomes

$$x_{t+1} - x_t = bx_t - x_t$$

or simply

$$x_{t+1} = bx_t. \quad (3.3)$$

That is, the population size at time $t + 1$ is the per capita birth rate b times the population size x_t at time t.

Suppose the number of individuals at time $t = 0$ is $x_0 = p \geq 0$. The statement $x_0 = p$ is called the *initial value* or *initial condition* for equation (3.3). The population size is therefore modeled by the *initial value problem* corresponding to assumptions (A1)–(A3):

$$
\begin{aligned}
x_{t+1} &= b x_t \\
x_0 &= p.
\end{aligned}
$$

The difference equation $x_{t+1} = b x_t$ is called *linear* because the state variable x appears linearly in the equation.

Suppose a petri dish culture initially consists of ten cells, and that each cell divides into two daughter cells every hour. Then $p = 10$ and $b = 2$, so the culture can be modeled by the initial value problem

$$
\begin{aligned}
x_{t+1} &= 2 x_t \\
x_0 &= 10,
\end{aligned}
$$

where the time step is one hour. Note that

$$
\begin{aligned}
x_0 &= 10 = 2^0 \, (10) \\
x_1 &= 2 x_0 = 2^1 \, (10) \\
x_2 &= 2 x_1 = 2 \, [2 \, (10)] = 2^2 \, (10) \\
x_3 &= 2 \, (2 \, [2 \, (10)]) = 2^3 \, (10) \\
&\vdots \\
x_t &= 2^t \, (10) \, .
\end{aligned}
$$

The function $x_t = 10 \, (2^t)$ is called the *closed-form solution* of the initial value problem. "Closed form" means that, given any value of t, you can compute the population size x_t without iterating. Here, we see an example of the fact that *linear dynamical systems have exponentially growing solutions*. Systems that grow exponentially are also said to grow *geometrically*, or to exhibit *Malthusian growth*.

In general, the solution of the initial value problem

$$
\begin{aligned}
x_{t+1} &= b x_t \\
x_0 &= p
\end{aligned}
\qquad (3.4)
$$

is (Exercise 4)

$$x_t = pb^t. \tag{3.5}$$

Note that the extinction state $x = 0$ is an equilibrium of model (3.4), since it satisfies the equilibrium equation

$$x_e = bx_e.$$

That is, if the initial condition is $x_0 = p = 0$, then solution (3.5) is the constant solution $x_t = 0$ for all t.

Now suppose $p > 0$. In this case, the behavior of solution (3.5) depends on the value of b. If $b > 1$, the population grows exponentially as b^t grows without bound. Thus, the equilibrium state $x_e = 0$ is unstable. If $b = 1$, the population remains at its initial size p for all time. Every value of x is an equilibrium solution in this case, and all the equilibria are neutrally stable. If $0 < b < 1$, the population declines exponentially toward zero as b^t declines, and so the equilibrium $x_e = 0$ is asymptotically stable.

You can see that the most important expression in model (3.4) is the number b; the value of b determines the long-term fate of the population. The parameter b is called the *intrinsic growth rate*. It is also called the *eigenvalue*.

Let x_t be the number of individuals in a population at time t. Assume

(A1) The average per capita birth rate is a constant $b > 0$ offspring per individual per unit time.

(A2) Generations are overlapping.

(A3) The average per capita death rate is a constant d deaths per individual per unit time, with $0 < d < 1$.

(A4) The only flows into or out of the population are due to births and deaths.

Then the population birth rate is bx_t offspring per unit time and the population death rate is dx_t deaths per unit time. Thus,

$$\begin{aligned} \Delta x_{t+1} &= \text{Birth rate at time } t - \text{Death rate at time } t \\ &= bx_t - dx_t \end{aligned}$$

or

$$x_{t+1} - x_t = bx_t - dx_t.$$

We can write the model in several equivalent forms to aid in various interpretations. For example, we can write

$$
\begin{aligned}
x_{t+1} &= x_t + bx_t - dx_t \\
\text{New census} &= \text{Last census} + \text{Births} - \text{Deaths}.
\end{aligned}
$$

We can also write

$$
\begin{aligned}
x_{t+1} &= bx_t + (1 - d)\, x_t \\
\text{New census} &= \text{Recruits} + \text{Survivors}.
\end{aligned}
$$

The term bx_t is called the recruitment term. The number $1 - d$ is called the survivorship, the probability that an individual will survive one unit of time. We can also write the model as

$$x_{t+1} = (b + 1 - d)\, x_t,$$

that is,

$$x_{t+1} = rx_t,$$

where we define the new parameter r to be $r = b + 1 - d$. Note that $r > 0$ since $b > 0$ and $d < 1$. This example is completed in Exercise 5.

3.4   NONLINEAR MAPS

What if a lack of resources due to overcrowding were to affect the birth rate? Let x_t be the size of a population at time t. Assume

(A1) The average per capita birth rate is a constant $b > 0$ offspring per individual per unit time at low populations sizes (i.e., if there are no *crowding effects*).

(A2) Because of crowding effects, the average per capita birth rate is reduced by the factor e^{-cx_t}, where $c > 0$ quantifies the strength of the crowding effect. (Note that $0 < e^{-cx_t} \le 1$.)

(A3) Generations are nonoverlapping.

By assumptions (A1) and (A2), the per capita birth rate in the presence of x_t individuals is be^{-cx_t}; therefore, the population birth rate is $be^{-cx_t}x_t$ offspring per unit time, and the model is

$$x_{t+1} = bx_te^{-cx_t} \qquad (3.6)$$

$$x_0 = p$$

with parameters $b, c > 0$. Model (3.6) is the famous *Ricker model*, historically used in fisheries (Ricker 1954; Ricker 1975). It is called *nonlinear* because the state variable x appears in a nonlinear way. This simple deterministic model can have incredibly complex dynamics, as we shall see in Chapter 4.

Let's find the equilibrium states of the Ricker model. The equilibrium equation is

$$x_e = bx_ee^{-cx_e}.$$

Note that $x_e = 0$ is a equilibrium solution. Suppose $x_e \neq 0$. If we divide both sides by x_e, we have

$$1 = be^{-cx_e},$$

which yields a nontrivial equilibrium solution $x_e = (\ln b)/c$.

It is often useful to graph the equilibria as a function of one of the parameters. In this case, we graph the two equilibria $x_e = 0, (\ln b)/c$ as a function of b (Figure 3.2). Note that the nontrivial equilibrium $x_e = (\ln b)/c$ is positive if and only if $b > 1$. The value $b = 1$ at which the two equilibrium branches cross is called a *bifurcation value* or *bifurcation point*. The graph of x_e vs b (Figure 3.2) is called a *bifurcation diagram*, and b is called the *bifurcation parameter*.

A model such as $x_{t+1} = bx_te^{-cx_t}$ is really a collection of infinitely many models, one for each specific pair of values of b and c. When looking at a bifurcation diagram such as Figure 3.2, one must remember that any particular system has a fixed value of b, for example, $b = 1.5$. Its equilibria are given by the values of the equilibrium branches directly above that specific value of b. Another way to say it is that any particular system "lives on a vertical line" in Figure 3.2.

What are the stabilities of the two equilibria $x_e = 0, (\ln b)/c$ of the Ricker map? Do their stabilities depend on the value of b? Clearly, we need to find a way to quantify and study stability in nonlinear maps. To this end, we now turn to one of the most important subjects in applied mathematics, the subject of *linearization*.

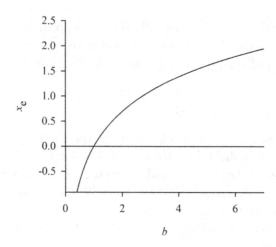

Figure 3.2: Bifurcation diagram for the Ricker map shows the equilibria x_e as a function of the parameter b. The two equilibrium branches are $x_e = 0$ and $x_e = (\ln b)/c$. Here, c = 1.

3.5 LINEARIZATION

3.5.1 Linearization of Functions

Suppose $f(x)$ is a function whose graph passes through the point $(a, f(a))$. Close to the point $(a, f(a))$, we can approximate the graph of f with its tangent line, that is, the line that is tangent to the graph of f at the point $(a, f(a))$. What is the equation of this line? Recall that the slope m of a line is the "rise over run." Given a nearby point (x, y) on the line (Figure 3.3), the rise from $(a, f(a))$ to (x, y) is $y - f(a)$, while the run is $x - a$. The slope is therefore

$$m = \frac{y - f(a)}{x - a},$$

and so

$$y - f(a) = m(x - a),$$

which is the "point-slope" formula for a line from high school algebra.

From calculus, we know that the slope m of the line tangent to f at the point $(a, f(a))$ is the derivative of f evaluated at $x = a$:

$$m = \frac{df}{dx}(a).$$

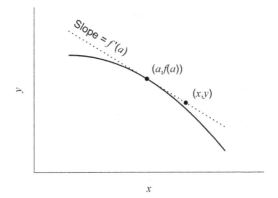

Figure 3.3: The equation of the tangent line at point $(a, f(a))$ is $y = f(a) + f'(a)(x - a)$.

Thus, the equation of the tangent line is

$$y - f(a) = \left[\frac{df}{dx}(a)\right](x - a),$$

or

$$y = f(a) + \left[\frac{df}{dx}(a)\right](x - a). \tag{3.7}$$

Equation (3.7) is called the *linearization of f at $x = a$*. It is also the equation of the tangent line.

Because the linearization is a good approximation of the function f near the point $x = a$, we have

$$f(x) \approx f(a) + \left[\frac{df}{dx}(a)\right](x - a) \text{ for } x \approx a.$$

Note that *you always linearize a function about a point*. It doesn't make sense to ask what the linearization of a function is without reference to a specific point $x = a$. Furthermore, *the linearization of a function is always a linear equation* of the form $y = mx + b$, where m and b are constants.

Linearization is one of the most important tools in applied mathematics and is extremely important in modeling. In fact, linearization provides the key to studying stability, as we shall soon see.

Let's find the linearization of the curve $f(x) = x^3 - 4x^2 + 7x - 1$ at $x = 2$. In this problem, we have $a = 2$. The derivative of f is

$$\frac{df}{dx} = 3x^2 - 8x + 7.$$

Thus,

$$f(a) = f(2) = 2^3 - 4(2)^2 + 7(2) - 1 = 5$$

and

$$\frac{df}{dx}(a) = \frac{df}{dx}(2) = 3(2)^2 - 8(2) + 7 = 3.$$

Hence, the linearization of f at $x = 2$ is

$$\begin{aligned}
y &= f(2) + \left[\frac{df}{dx}(2)\right](x - 2) \\
&= 5 + 3(x - 2) \\
&= 3x - 1,
\end{aligned}$$

and therefore, we have the linear approximation

$$f(x) \approx 3x - 1 \text{ for } x \approx 2,$$

that is,

$$x^3 - 4x^2 + 7x - 1 \approx 3x - 1 \text{ for } x \approx 2,$$

which is a neat result.

3.5.2 Linearization of Discrete-Time Maps

Consider the nonlinear discrete dynamical system

$$x_{t+1} = f(x_t) \tag{3.8}$$

with equilibrium x_e. Note that the constant x_e is an equilibrium of the difference equation $x_{t+1} = f(x_t)$ if and only it satisfies the equilibrium equation $x_e = f(x_e)$. When we say equation (3.8) is *nonlinear*, we mean that the right-hand side f is a nonlinear function of x_t. An example of such an equation would be

$$x_{t+1} = ax_t^2.$$

What we would like to do is to replace the nonlinear $f(x_t)$ on the right-hand side of equation (3.8) with a simple linear function that approximates it for values of $x_t \approx x_e$.

In the previous section, we learned that

$$f(x) \approx f(a) + \left[\frac{df}{dx}(a)\right](x - a) \text{ for } x \approx a.$$

In the current context, x is x_t, and a is x_e. Making those substitutions, we have

$$f(x_t) \approx f(x_e) + \left[\frac{df}{dx}(x_e)\right](x_t - x_e) \text{ for } x_t \approx x_e.$$

Also, since x_e is an equilibrium of equation (3.8), we know that $x_e = f(x_e)$.

Thus, we can write

$$f(x_t) \approx x_e + \left[\frac{df}{dx}(x_e)\right](x_t - x_e) \text{ for } x_t \approx x_e.$$

It is traditional to define the Greek letter lambda (λ) to be the derivative df/dx evaluated at x_e:

$$\lambda = \frac{df}{dx}(x_e).$$

Note that λ is simply a constant, a number that you can compute. It is called the *eigenvalue*.

Then

$$\begin{aligned} x_{t+1} &= f(x_t) \\ &\approx x_e + \lambda(x_t - x_e) \text{ for } x_t \approx x_e, \end{aligned}$$

and hence

$$x_{t+1} - x_e \approx \lambda(x_t - x_e) \text{ for } x_t \approx x_e. \tag{3.9}$$

We can restate equation (3.9) in words: As long as the system is in the neighborhood of the equilibrium, the displacement of the system state from equilibrium at time $t + 1$ is approximately the number λ times the displacement of the system from equilibrium at time t. That is, x_{t+1} is about λ times as far from the equilibrium as x_t was. You can see that if $-1 < \lambda < 1$, this is good news for the stability of equilibrium x_e, because the distance between the system state and the equilibrium is shrinking as time goes on—in fact, it appears that the equilibrium x_e would be asymptotically stable. If $\lambda > 1$ or $\lambda < -1$, however, the distance between the system state and the equilibrium grows whenever the system is near the equilibrium, and so it appears that x_e would be unstable.

We now clean up the notation a bit and state these observations in a theorem. Define the displacement, or variation, of the system state from equilibrium to be

$$z_t = x_t - x_e. \tag{3.10}$$

Using this as a change of variables, we can rewrite equation (3.9) as

$$z_{t+1} \approx \lambda z_t \text{ for } z_t \approx 0.$$

Note that the change of variables (equation 3.10) simply shifts the equilibrium to zero.

The equation

$$z_{t+1} = \lambda z_t \text{ for } z_t \approx 0$$

is called the *variation equation*. The variation equation approximates the dynamic change in displacement z when the system is close to its equilibrium. Note that *the variation equation is linear*, and its solution is

$$z_t = z_0 \lambda^t.$$

Thus, *the displacement of the population from equilibrium grows or decays exponentially when the population size is near its equilibrium value.* Whether the displacement grows or decays depends on whether the eigenvalue λ is greater than or less than one in absolute value, that is, whether $|\lambda| > 1$ or $|\lambda| < 1$.

Definition 3.1 *The* **linearization** *of* $x_{t+1} = f(x_t)$ *at the* **equilibrium** x_e *is the linear map*

$$z_{t+1} = \lambda z_t,$$

where the **eigenvalue** λ *is given by*

$$\lambda = \frac{df}{dx}(x_e).$$

Note that *you must always linearize a difference equation about a specific equilibrium*. It does not make sense to "find the linearization" without reference to a particular equilibrium. Furthermore, *the linearization of a nonlinear difference equation is always a linear difference equation of the form* $z_{t+1} = \lambda z_t$, *where λ is a number*.

Definition 3.2 *An equilibrium x_e of $x_{t+1} = f(x_t)$ is* **hyperbolic** *if and only if $|\lambda| \neq 1$.*

Theorem 3.1 *(Linearization Theorem) Let $x_{t+1} = f(x_t)$ be a nonlinear map with hyperbolic equilibrium x_e. Suppose the function f is continuously differentiable in x (that is, df/dx exists and is continuous). Then*

$$|\lambda| < 1 \Longrightarrow x_e \text{ is asymptotically stable}$$
$$|\lambda| > 1 \Longrightarrow x_e \text{ is unstable,}$$

where

$$\lambda = \frac{df}{dx}(x_e).$$

Note that the *nonhyperbolic case* $|\lambda| = 1$ *is not covered by the linearization theorem*. If $|\lambda| = 1$, the theorem simply does not apply, and no conclusion can be drawn from the theorem.

3.5.3 Linearizing the Ricker Map

Consider again the Ricker map

$$x_{t+1} = bx_t e^{-cx_t},$$

where $b, c > 0$. Here, $f(x) = bxe^{-cx}$. The equilibria are

$$x_e = 0 \text{ and } x_e = \frac{\ln b}{c}.$$

Let's linearize the Ricker map about each of its equilibria. In both cases, we will need the derivative of f with respect to x:

$$\frac{df}{dx} = be^{-cx}(1 - cx).$$

We first linearize about $x_e = 0$. Now,

$$\lambda = \frac{df}{dx}(0) = be^0(1 - 0) = b,$$

and so the linearization at $x_e = 0$ is

$$z_{t+1} = bz_t.$$

Recalling that $b > 0$, we have

$$0 < b < 1 \implies x_e = 0 \text{ is asymptotically stable}$$
$$b > 1 \implies x_e = 0 \text{ is unstable.}$$

Now we linearize about the nontrivial equilibrium $x_e = \frac{\ln b}{c}$:

$$
\begin{aligned}
\lambda &= \frac{df}{dx}\left(\frac{\ln b}{c}\right) \\
&= be^{-c\left(\frac{\ln b}{c}\right)}\left(1 - c\left(\frac{\ln b}{c}\right)\right) \\
&= be^{\ln b^{-1}}(1 - \ln b) \\
&= bb^{-1}(1 - \ln b) \\
&= 1 - \ln b.
\end{aligned}
$$

The linearization at $x_e = \frac{\ln b}{c}$ is therefore

$$z_{t+1} = (1 - \ln b)z_t,$$

and so

$$|1 - \ln b| < 1 \implies x_e = \frac{\ln b}{c} \text{ is asymptotically stable} \quad (3.11)$$

$$|1 - \ln b| > 1 \implies x_e = \frac{\ln b}{c} \text{ is unstable.}$$

In order to obtain simple conditions on b for stability, we need to do the algebra necessary to remove the absolute values. Recall that

$$|x| < p \text{ if and only if } -p < x < p \quad (3.12)$$

$$|x| > p \text{ if and only if } x < -p \text{ or } x > p.$$

From inequalities (3.11) and (3.12), we obtain (Exercise 1):

$$1 \; < \; b < e^2 \; \implies \; x_e = \frac{\ln b}{c} \text{ is stable} \tag{3.13}$$

$$0 \; < \; b < 1 \; \implies \; x_e = \frac{\ln b}{c} \text{ is unstable}$$

$$b \; > \; e^2 \; \implies \; x_e = \frac{\ln b}{c} \text{ is unstable.}$$

We can now indicate stability on our bifurcation diagram (Figure 3.4).

Note that if $b > e^2$, both equilibria are unstable! How do the solutions of the Ricker map behave for these values of b? Toward what value do they tend, if any? What is the long-term fate of the system? We will investigate this fascinating question in Chapter 4. Before moving on, however, let's look more carefully at the Ricker nonlinearity. This type of nonlinearity shows up a lot in models of population dynamics. We might ask: Is it simply phenomenological and descriptive, or is there a mechanistic underpinning?

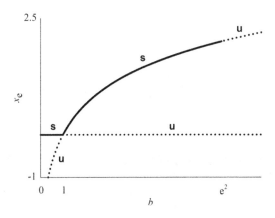

Figure 3.4: Bifurcation diagram for the Ricker map showing stability of equilibria. Dotted equilibrium branches are unstable (u); solid are stable (s). Here, c = 1.

3.6 THE RICKER NONLINEARITY

The Ricker nonlinearity e^{-cx} is typically a result of random deleterious contacts between individuals in a population. For example, in fisheries or insect models, the Ricker function can result from random cannibalistic encounters between individuals within a population.

Suppose, for example, that we are modeling an insect population with J juveniles and A adults using a discrete-time model. Suppose the adults cannibalize juveniles through random encounters. Let Δt be a small fraction of one model time step, say one-mth of the time step, so that $1 = m\Delta t$. Assume that the probability that one adult encounters (and hence cannibalizes) one juvenile during Δt units of time is proportional to the time elapsed Δt, that is, $c\Delta t$. The coefficient $c > 0$ can be thought of as the continuous-time "instantaneous" encounter rate or cannibalism rate. The probability that one juvenile survives cannibalism by one adult in Δt units of time is therefore $1 - c\Delta t$. The probability that a juvenile survives all of the A adults in Δt units of time (assuming the encounters are independent events) is the product

$$\prod_{i=1}^{A} (1 - c\Delta t) = (1 - c\Delta t)^A .$$

Now let's consider what happens during one time step of the model. This time period is composed of $m = 1/\Delta t$ successive times of duration Δt. The probability that a juvenile survives all of the adults during this time is the product

$$\prod_{i=1}^{m} (1 - c\Delta t)^A = (1 - c\Delta t)^{Am}$$
$$= (1 - c\Delta t)^{A/\Delta t} .$$

If we let $\Delta t \to 0$, we find that (Exercise 2)

$$\lim_{\Delta t \to 0} (1 - c\Delta t)^{A/\Delta t} = e^{-cA}. \tag{3.14}$$

So, the probability that a juvenile survives cannibalism during one model time step in the presence of A adults is e^{-cA}, which is the Ricker nonlinearity.

3.7 EXERCISES

1. Use inequalities (3.11) and (3.12) to obtain inequalities (3.13), keeping in mind that $b > 0$.

2. Prove equation (3.14). Hint: You will need L'Hospital's rule.

3. A population of annual plants has an average per capita fecundity of 120.3 seeds per plant per year. Approximately 12% of these seeds germinate. About 10% of the resulting seedlings survive until maturity. The initial population census (year $t = 0$) is 52 plants.

   a. How can the fecundity be 120.3? That is, what does 0.3 seeds mean?

   b. Write down the population model (the initial value problem).

   c. Find the closed-form solution of the initial value problem.

   d. Will the population grow or decline? Explain.

   e. Is the extinction state stable or unstable?

   f. According to the model, how many plants will there be at year $t = 5$?

   g. How many years will it take for the population to grow to 75 plants, according to the model?

   h. How many years will it take for the population to double its initial size, according to the model? This is called the *doubling time*.

   i. Explain why the doubling time in this model does not depend on the initial population size.

4. Consider the linear model

$$x_{t+1} = bx_t \qquad (3.15)$$
$$x_0 = p.$$

   a. Use mathematical induction (MI) to prove that the closed-form solution is $x_t = pb^t$.

b. If $p > 0$ and $0 < b < 1$, what is $\lim_{t \to \infty} x_t$? What does this mean in biological terms?

c. If $p > 0$ and $b > 1$, what is $\lim_{t \to \infty} x_t$? What does this mean in biological terms?

d. If $p > 0$ and $b = 1$, what is $\lim_{t \to \infty} x_t$? What does this mean in biological terms?

e. Write a program that *iterates* recursion formula (3.15) to produce a time series of length n. In your program, set $p = 50$ and $n = 5$. Graph the time series for each of $b = 0.5$, 0.9, 1.0, 1.1, and 1.5. Display all five time series on the same graph. Turn in the program and the graph.

5. Consider the linear population model

$$\begin{aligned} x_{t+1} &= (b + 1 - d)x_t \\ x_0 &= p > 0, \end{aligned}$$

where $b > 0$ and $0 < d < 1$.

a. Give the closed-form solution.

b. Use the closed-form solution to explain why

$$\begin{aligned} b &< d \implies \lim_{t \to \infty} x_t = 0 \\ b &> d \implies \lim_{t \to \infty} x_t = \infty \\ b &= d \implies x_t = p \text{ for all } t. \end{aligned}$$

c. Give a biological interpretation for the results in (b). Does this make sense biologically?

6. Find the linearization of the function $f(x) = x^4 - x + 15$ at $x = 1$. Graph the function and its linearization on the same axes near $x = 1$.

7. Find the linearization of the function $f(x) = \sin x$ at $x = 0$. Graph the function and its linearization on the same axes near $x = 0$.

8. In this problem, you will carry out a complete equilibrium stability analysis for the *Beverton-Holt population model*. This model was introduced by Beverton and Holt in 1957 in the context of fisheries (Beverton and Holt 1957):

$$
\begin{aligned}
x_{t+1} &= \frac{bx_t}{1 + cx_t} \\
x_0 &= p \geq 0 \\
b, c &> 0.
\end{aligned} \tag{3.16}
$$

a. Is the Beverton-Holt model (3.16) linear or nonlinear? Why?

b. Find all the equilibria.

c. Graph the equilibria x_e as functions of the parameter b.

d. Under what conditions is the nontrivial equilibrium positive?

e. Find the linearization of the Beverton-Holt model at each equilibrium.

f. For what values of b is each equilibrium stable/unstable?

g. Indicate the stabilities on your bifurcation diagram in (c).

h. What are the bifurcation points on your diagram in (c)?

i. Write a program that iterates the Beverton-Holt map to produce a time series of length n. In your program, set $p = 15$, $c = 0.002$, and $n = 100$. Graph the time series for each of $b = 0.5$, 0.9, 1.0, 1.1, and 1.5. Display all five time series on the same graph. Turn in the program and the graph.

9. In this problem, you will carry out a complete equilibrium stability analysis for the so-called *discrete logistic map*

$$
\begin{aligned}
x_{t+1} &= rx_t (1 - x_t) \\
0 &< x_0 < 1 \\
0 &< r < 4.
\end{aligned} \tag{3.17}
$$

Here, x_t is the density of organisms (number of individuals per unit area or volume) at time t.

a. Is the discrete logistic map (3.17) linear or nonlinear? Why?

b. Find all the equilibria.

c. Graph the equilibria x_e as functions of the parameter r.

d. Under what conditions is the nontrivial equilibrium positive?

e. Find the linearization of the model at each equilibrium.

f. For what values of r is each equilibrium stable/unstable?

g. Indicate the stabilities on your bifurcation diagram in (c).

h. What are the bifurcation points on your diagram in (c)?

i. Write a program that iterates the discrete logistic map to produce a time series of length n. In your program, set $x_0 = 0.5$ and $n = 30$. Graph the time series for each of $r = 0.5$, 1.5, 2.5, 3.2, and 3.5. Display all five time series on the same graph. Turn in the program and the graph. Also turn in the five numerical time series as lists of numbers.

j. Explore the behavior of the discrete logistic map for $3.5 < r < 4$. You may have to increase r by small increments to see how the time series change.

k. Mathematically, what happens to solutions if $r > 4$? Hint: By hand, iterate $x_{t+1} = 5x_t(1 - x_t)$ starting with $x_0 = 0.5$ to see what happens.

BIBLIOGRAPHY

Beverton, R. J. H. and Holt, S. J. 1957. On the dynamics of exploited fish populations. *Fishery Investigations (II)* 19:1–533. [Introduces the Beverton-Holt mathematical model.]

Ricker, W. E. 1954. Stock and recruitment. *Journal of the Fisheries Research Board of Canada* 11:559–623. DOI:10.1139/f54-039. [Classic paper introducing the Ricker model.]

Ricker, W. E. 1975. Computation and interpretation of biological statistics of fish populations. *Bulletin of the Fisheries Research Board of Canada*, No. 119, Ottawa.

CHAPTER 4

Chaos: Simple Rules Can Generate Complex Results

4.1 WHAT YOU SHOULD KNOW ABOUT THIS CHAPTER

In this chapter, we return to the Ricker map from Chapter 3 and explore its dynamics beyond the point at which all the equilibria destabilize. To complete the exercises in this chapter, you will need a program that plots bifurcation diagrams for discrete maps. An example of such a program is the freeware E&F Chaos (see bibliography). Alternately, you can write your own code (Exercise 1).

Chaos is a fascinating subject that has captured the interest of the general public as well as that of scientists and mathematicians. If you would like to read more about chaos, check out the popular book by James Gleick (Gleick 1987).

4.2 RICKER MODEL REVISITED

In Chapter, 3 we studied the equilibria $x_e = 0$ and $x_e = c^{-1} \ln b$ of the Ricker map

$$
\begin{aligned}
x_{t+1} &= b x_t e^{-c x_t} \\
x_0 &\geq 0 \\
b, c &> 0.
\end{aligned}
$$

DOI: 10.1201/9781003265382-6

We used the linearization theorem to prove that

$$0 \; < \; b < 1 \Longrightarrow x_e = 0 \text{ is stable}$$
$$b \; > \; 1 \Longrightarrow x_e = 0 \text{ is unstable}$$

and that

$$1 \; < \; b < e^2 \Longrightarrow x_e = \frac{\ln b}{c} \text{ is stable}$$
$$0 \; < \; b < 1 \Longrightarrow x_e = \frac{\ln b}{c} \text{ is unstable}$$
$$b \; > \; e^2 \Longrightarrow x_e = \frac{\ln b}{c} \text{ is unstable.}$$

We summarized this information in a bifurcation diagram, shown again here as Figure 4.1.

The equilibrium analysis of the Ricker map begs the question, "What happens to solutions if $b \; > \; e^2$?" For these values of b, nonequilibrium solutions cannot equilibrate, for there are no stable equilibria for them to approach. Some exploratory computer simulations are in order.

Let $c = 0.01$ and $x_0 = 75$. Let's set b at various values and inspect the resulting time series. Figure 4.2 shows the time series generated for $b = 0.5$, $b = 1.3$, $b = 3.6$, and $b = 5.7$. If $b = 0.5$, solutions

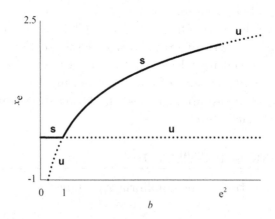

Figure 4.1: Bifurcation diagram of the Ricker map showing the stability of equilibria. Dotted equilibria are unstable (u); solid are stable (s). Here, c = 1.

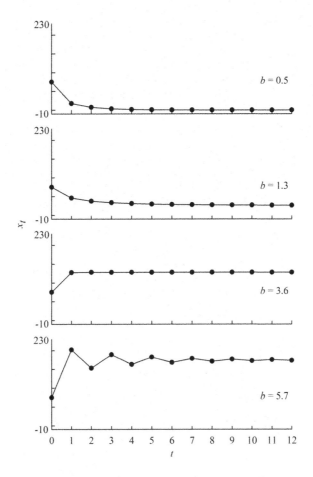

Figure 4.2: Time series for the Ricker map with $b = 0.5$, 1.3, 3.6, 5.7. In each case, $c = 0.01$.

equilibrate to zero, whereas if $b = 1.3$, solutions equilibrate to a positive equilibrium $x_e \approx 26.3$. The *attractors* in these cases are the sets $\{0\}$ and $\{26.3\}$, respectively. For $b = 3.6$, the equilibrium is higher, about $x_e \approx 128.1$, and for $b = 5.7$, the equilibrium value is even higher, about $x_e \approx 174.0$. The attractors are the sets $\{128.1\}$ and $\{174.0\}$, respectively. The interesting thing in this last case is that solutions oscillate as they approach the equilibrium.

Figure 4.3 shows time series for $b = 8$, $b = 14$, $b = 14.6$, and $b = 17$. We know from our previous linearization calculations that for

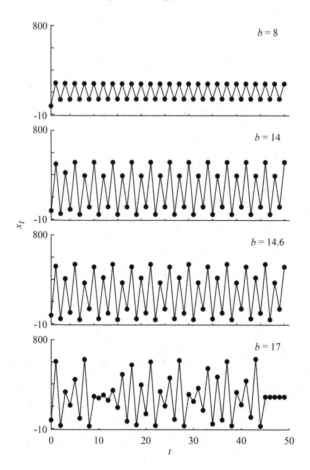

Figure 4.3: Ricker time series for $b = 8, 14, 14.6, 17$. In each case, $c = 0.01$.

$b > e^2 \approx 7.389$, the solutions no longer equilibrate. The computer simulation for $b = 8$ in Figure 4.3 suggests solutions approach a *two-cycle*; that is, they begin to oscillate between the two values $x \approx 138.6$ and $x \approx 277.3$. The attractor in this case is the set of two points $\{138.6, 277.3\}$. Further numerical explorations near $b = e^2$ suggest that as b is increased through the value e^2, the point attractor (for $b < e^2$) bifurcates into two points (for $b > e^2$). The value $b = e^2$ is called a *bifurcation value*. For $b = 14$, solutions approach a *four-cycle* attractor, and for $b = 14.6$, an *eight-cycle*. As b is increased, the periods of these cycles continue to double, until at a critical value of

b, there is a transition to an aperiodic, complicated dynamic called
chaos. This sequence of bifurcations is called a *period-doubling cascade
to chaos*.

If the attractor, also called the *final state*, of each time series is
plotted against the parameter b, we can continue the Ricker bifurcation
diagram in Figure 4.1 beyond $b = e^2$ (Figure 4.4). Note that unstable
equilibria and cycles are not shown in Figure 4.4, since computer
iterations will only identify the stable ones. Chaos has a complicated
mathematical definition which we will not address in this book. For
our purposes, here we simply note three important characteristics
of chaos. First, *chaos is deterministic*. Note that the chaotic time
series shown at the bottom of Figure 4.3 is completely determined
by the simple iterative rule $x_{t+1} = 17x_t e^{-0.01x_t}$, where $x_0 = 75$.
Second, chaotic dynamics, although deterministic, "look" random.
Third, chaotic dynamics exhibit *sensitivity to initial conditions*. In
Figure 4.5, we see two time series of the Ricker model with $b = 17$
and $c = 0.01$. The solid curve begins with the initial condition $x_0 =
75$, whereas the dotted curve is generated from the initial condition
$x_0 = 75.1$. The two time series are close together for several time steps,
but soon diverge and look nothing like each other. This is sensitivity to
initial conditions: A small change or error in initial conditions is quickly
magnified.

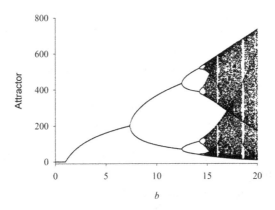

Figure 4.4: Bifurcation diagram for the Ricker map with $c = 0.01$.

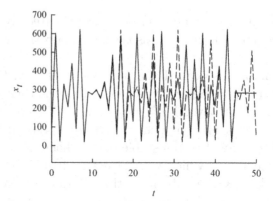

Figure 4.5: Two time series of the Ricker map with $b = 17$ and $c = 0.01$. The solid curve has initial condition $x_0 = 75$, and the dashed curve has initial condition $x_0 = 75.1$. The two solutions soon diverge, showing sensitivity to initial conditions.

4.3 NEW PARADIGMS ARISE FROM CHAOS

4.3.1 Deterministic Unpredictability

Determinism has generally been equated with *predictability*. If a system is deterministic, then the output is completely determined by the input and is thus predictable, given the input. Similarly, *stochasticity* has been identified with *unpredictability*, for obvious reasons.

Chaos, however, significantly alters this classical paradigm. Chaos is deterministic, but has sensitivity to initial conditions. Therefore, if the initial conditions are not known precisely (and they never are in the real world), then the future state of the system is, for all practical purposes, unpredictable. So in a very real sense, there is such a thing as "deterministic unpredictability."

4.3.2 Complex Results Can Arise from Simple Rules

Another classical paradigm has been that simple causes give rise to simple results, and (equivalently) complex results imply complex causes. Chaos shows emphatically that this is not necessarily the case. The Ricker model is a simple recursion formula, and yet its dynamics are so complicated that a mathematician can spend a lifetime exploring them. Thus, we see from this example that complex results do *not* necessary imply complex causes. Very simple rules can generate extraordinarily complicated results.

4.4   MAY'S HYPOTHESIS

In 1976, Robert May wrote (May 1976):

> Quite apart from their intrinsic mathematical interest, the above results raise very awkward biological questions. They show that simple and fully deterministic models, in which all biological parameters are exactly known, can nonetheless (if the nonlinearities are sufficiently severe) lead to population dynamics which are in effect indistinguishable from the sample function of a random process. Apparently chaotic population fluctuations need not necessarily be due to random environmental fluctuations, or sampling errors, but may reflect the workings of some deterministic, but strongly density dependent, population model.

May's hypothesis is the idea that the apparently random fluctuations of population abundances can be explained largely by low-dimensional, nonlinear, deterministic forces.

In 1997, chaos was documented in population dynamics for the first time when it was experimentally induced in laboratory populations of insects (Costantino et al. 1997; Cushing et al. 2003; Dennis et al. 2001). We will discuss these experiments in Chapter 6. To date, no one knows whether chaos occurs "naturally" in field populations; speculation and argumentation abound.

4.5   EXERCISES

In this set of exercises, you will need a program that draws bifurcation diagrams for discrete maps. An example is the freeware program E&F Chaos (see bibliography entry for URL). Alternately, you can write your own code (Exercise 1).

1. Write your own code to reproduce Figure 4.4, which is the bifurcation diagram for the Ricker map. Hint: You will need to

specify a "mesh" of values of the bifurcation parameter b along the horizontal axis. For each of these fixed values of b, you will iterate the map a fixed number of times, maybe 300, so that the time series approaches the attractor. The early iterates, before the time series is close to the attractor, are called "transients," and you will not plot them in the bifurcation (final state) diagram. You will plot the later iterates, say iterates 200–300, as a scatter plot above the associated value of b.

2. Use the computer to draw a bifurcation diagram for the *Beverton-Holt model*

$$x_{t+1} = \frac{bx_t}{1 + cx_t}$$
$$x_0 > 0$$
$$b, c > 0$$

using b as the bifurcation parameter. Set $c = 1$, and let $0 \le b \le 10$.

3. Use the computer to draw a bifurcation diagram for the so-called *discrete logistic model*

$$x_{t+1} = bx_t (1 - x_t)$$
$$0 < x_0 < 1$$
$$0 < b < 4$$

using b as the bifurcation parameter. What happens to solutions if $b > 4$?

4. Consider the *Ricker model with survivorship*:

$$x_{t+1} = bx_t e^{-cx_t} + (1 - \mu) x_t$$
$$x_0 > 0$$
$$b, c > 0$$
$$0 \le \mu \le 1.$$

Here, μ is the fraction of the population that does not survive one time step and $1 - \mu$ is the fraction that does survive one time step.

a. Set $c = 0.01$. Use the computer to draw bifurcation diagrams for each of the following values of μ, using b as the bifurcation parameter with $0 \le b \le 100$.

   i. $\mu = 1.0$ (Ricker map without survivorship)

   ii. $\mu = 0.9$

   iii. $\mu = 0.8$

   iv. $\mu = 0.7$

   v. $\mu = 0.6$

   vi. $\mu = 0.5$

   vii. $\mu = 0.4$

   viii. $\mu = 0.3$

   ix. $\mu = 0.2$

   x. $\mu = 0.1$

   xi. $\mu = 0.0$

b. Set $c = 0.01$ and $b = 80$. Use the computer to draw the bifurcation diagram using μ as the bifurcation parameter with $0 \le \mu \le 1$. Relate this diagram to the sequence of diagrams in part (4a).

5. In this problem, you will use the computer to further explore the bifurcation diagram of the Ricker model

$$x_{t+1} = bx_t e^{-cx_t}$$
$$x_0 > 0$$
$$b, c > 0$$

using b as the bifurcation parameter. Let $c = 1$.

a. Draw the bifurcation diagram for $0 \le b \le 50$.

b. Draw the bifurcation diagram for $21 \le b \le 26$.

c. Draw the bifurcation diagram for $23 \le b \le 25$.

d. Draw the bifurcation diagram for $24.8 \le b \le 25$. Zoom in on the neighborhood of $0.3 \le x \le 1.8$ and $24.88 \le b \le 24.95$. This illustrates the *fractal* nature of the bifurcation diagram, in which patterns are repeated at ever-smaller scales. Fractal structures are deeply related to chaos.

BIBLIOGRAPHY

Costantino, R. F., Desharnais, R. A., Cushing, J. M., and Dennis, B. 1997. Chaotic dynamics in an insect population. *Science* 275:389–391. DOI: 10.1126/science.275.5298.389. ["Announcement" paper heralding the documentation of chaos in a laboratory insect population. A follow-up paper gave the details. See Dennis et al. (2001) below. Both theoretical and empirical: models connected to data.]

Cushing, J. M., Costantino, R. F., Dennis, B., Desharnais, R. A., and Henson, S. M. 2003. *Chaos in Ecology: Experimental Nonlinear Dynamics.* Academic Press, San Diego. [Summarizes the large body of work involved in the documentation of chaos in a laboratory population. In Chapter 6 we will draw heavily on this book to showcase how dynamic population data can be connected to mathematical models.]

Dennis, B., Desharnais, R. A., Cushing, J. M., Henson, S. M., and Costantino, R. F. 2001. Estimating chaos and complex dynamics in an insect population. *Ecological Monographs* 71:277–303. DOI: 10.1890/0012-9615(2001)071[0277:ECACDI]2.0.CO;2. [Detailed follow-up to the announcement paper in *Science*. See Ref. Costantino et al. 1997]

E&F Chaos. https://cendef.uva.nl/software/ef-chaos/ef-chaos.html. [Useful tool for drawing bifurcation diagrams.]

Gleick, J. 1987. *Chaos: Making a New Science.* Viking, New York. [This popular book was a finalist for the Pulitzer Prize in 1987.]

May, R. M., ed. 1976. *Theoretical Ecology: Principles and Applications.* W. B. Saunders, Philadelphia. [Played a foundational role in the development of theoretical ecology.]

CHAPTER 5

Higher-Dimensional Discrete-Time Models

5.1 WHAT YOU SHOULD KNOW ABOUT THIS CHAPTER

In this chapter, we consider how to model discrete-time systems whose descriptions require more than one state variable. Before beginning this chapter, you should review basic matrix operations and learn (or review) some introductory linear algebra by working through the self-guided tutorial in Appendix A. A course in linear algebra is not a prerequisite for this textbook; everything you need to know from linear algebra for this and subsequent chapters is contained in Appendix A.

Also, note that in this chapter, you will need to know how to take *partial derivatives*, a topic encountered in third-semester calculus. For those who have not met partial derivatives, they are easy to learn. If you have a function of (say) two variables $f(x, y)$, then the partial derivative of f with respect to x, denoted $\partial f / \partial x$, is the derivative of f with respect to x while holding y constant. So, for example, if $f(x, y) = x^2 + y^2 + x^3 y^3$, then the partial derivatives are (showing zeros to make a point):

$$\frac{\partial f}{\partial x} = 2x + 0 + y^3 \left(3x^2\right)$$
$$\frac{\partial f}{\partial y} = 0 + 2y + x^3 \left(3y^2\right).$$

If this is your first experience with partial derivatives, work a few problems in a calculus book to get comfortable with them.

DOI: 10.1201/9781003265382-7

The main idea in this chapter is to learn how to solve and analyze linear systems, and then to apply those techniques to nonlinear systems through the method of linearization.

5.2 INTRASPECIFIC INTERACTIONS

Up to this point, we have modeled populations with one state variable, usually the number of organisms (or density of organisms per unit area or volume). In doing this, we have assumed that all members of the population are similar. For many populations, this is a false assumption. For example, flour beetles (*Tribolium*) undergo *complete metamorphosis* and, in the process, exhibit four discrete life stages— egg, larva, pupa, and adult. *Life-cycle* stages can have different durations, behaviors, *intraspecific interactions* (interactions within species), and *vital rates* (e.g., birth and death rates).

To model the population dynamics of organisms that pass through various life-cycle stages, we typically need to use more than one state variable. For example, flour beetles require at least three state variables: one for larvae, one for pupae, and another for adults. The egg stage can sometimes be ignored because of its relatively short duration (Cushing et al. 2003). The resultant model is called an *age-structured model*, or more generally, a *stage-structured model*. These models are used to describe the dynamics of populations whose vital rates and interactions depend on the ages or stages of the organisms. Discrete-time age-structured models are called *Leslie matrix models*, after biologist Patrick H. Leslie. The number of state variables is called the *dimension* of the system. A well-known standard text about Leslie matrix models is the one by Hal Caswell (Caswell 2001).

5.3 INTERSPECIFIC INTERACTIONS

When modeling a population, it is also important to consider whether any *interspecific interactions* (interactions between species) play a significant role in the dynamics. In terms of dynamics, the three main categories of interspecific interactions are *predator-prey*, *mutualism* (*cooperation*), and *competition* (Table 5.1). In predator-prey systems, the presence of the predator species has a negative effect on the prey species, while the prey species has a positive effect on the predator

TABLE 5.1   Species Interactions

	Effect of Species 1 on Species 2	Effect of Species 2 on Species 1
Predator-prey	−	+
Mutualism	+	+
Competition	−	−

species. Host-parasite systems also fit in this category. In mutualistic systems (cooperative systems), each species has a positive effect on the other. In competitive systems, each species has a negative effect on the other.

Models of interacting species require more than one state variable. For example, a multi-species system consisting of one predator and two prey species would require at least a three-dimensional model.

5.4   EXAMPLE OF AN AGE-STRUCTURED SINGLE-SPECIES MODEL

Consider an age-structured juvenile-adult model for a fish population, with the following assumptions:

(A1) An adult fish has probability μ_A of dying during one year, where $0 < \mu_A < 1$.

(A2) Adult fish deposit an average of $b > 0$ fertilized eggs per adult per year.

(A3) A fertilized egg hatches with probability $p > 0$.

(A4) A juvenile fish matures sexually in one year.

(A5) A juvenile fish has probability μ_J of dying before maturing, where $0 < \mu_J < 1$.

These assumptions are summarized in the *Leslie diagram* in Figure 5.1.

Let J_t and A_t be the number of juvenile and adult fish, respectively, at time t. The time step is one year, the duration of the juvenile stage.

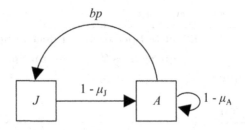

Figure 5.1: Leslie diagram for an age-structured fish population model showing per capita flow rates.

In general, we can write

$$
\begin{aligned}
J_{t+1} &= \text{Juvenile recruits} + \text{juveniles carried over from last census} \\
A_{t+1} &= \text{Adult recruits} + \text{adults surviving from last census.}
\end{aligned}
$$

From the assumptions, we know that

$$
\begin{aligned}
\text{Juvenile recruits} &= (\text{Eggs per adult})\,(\text{adults at time } t) \\
&\quad \times (\text{fraction eggs that hatch}) \\
&= bA_t p \\
\text{Juvenile carryovers} &= 0 \\
\text{Adult recruits} &= (\text{Juveniles at time } t) \\
&\quad \times (\text{fraction that survive until maturity}) \\
&= J_t\,(1 - \mu_J) \\
\text{Adult survivors} &= (\text{Adults at time } t) \\
&\quad \times (\text{fraction that survive one year}) \\
&= A_t\,(1 - \mu_A)\,.
\end{aligned}
$$

The Leslie model is therefore

$$
\begin{aligned}
J_{t+1} &= bpA_t \\
A_{t+1} &= (1 - \mu_J)\,J_t + (1 - \mu_A)\,A_t.
\end{aligned}
\tag{5.1}
$$

We can also write equation (5.1) as a matrix equation:

$$
\begin{pmatrix} J_{t+1} \\ A_{t+1} \end{pmatrix} = \begin{pmatrix} 0 & bp \\ (1 - \mu_J) & (1 - \mu_A) \end{pmatrix} \begin{pmatrix} J_t \\ A_t \end{pmatrix}.
$$

The matrix

$$M = \begin{pmatrix} 0 & bp \\ (1 - \mu_J) & (1 - \mu_A) \end{pmatrix}$$

is called a *Leslie matrix*, or *projection matrix*. System (5.1) is a two-dimensional linear model: two-dimensional because it has two state variables J_t and A_t, and linear because both equations in (5.1) are linear in the state variables J and A.

5.5   EXAMPLE OF A TWO-SPECIES MODEL

Consider two similar species with nonoverlapping generations, competing for space, with the following assumptions:

(A1) In the absence of crowding, species 1 and 2 have per capita birth rates of b_1 and b_2, respectively.

(A2) Each species suffers fractional reductions in per capita birth rate due to crowding by its own species and by the other species. Each fractional reduction is of Ricker type, that is, of the form e^{-cN}, where N is the number of competing organisms.

Let x_t and y_t be the numbers of species 1 and 2, respectively, at time t. Then

$$\begin{aligned} x_{t+1} &= b_1 x_t e^{-c_{11} x_t} e^{-c_{12} y_t} \\ y_{t+1} &= b_2 y_t e^{-c_{21} x_t} e^{-c_{22} y_t}, \end{aligned}$$

or

$$\begin{aligned} x_{t+1} &= b_1 x_t e^{-c_{11} x_t - c_{12} y_t} \\ y_{t+1} &= b_2 y_t e^{-c_{21} x_t - c_{22} y_t}. \end{aligned} \qquad (5.2)$$

The system (5.2) is a two-dimensional nonlinear model: two-dimensional because it has two state variables x_t and y_t, and nonlinear because the right-hand sides of the equations in (5.2) are nonlinear functions of the state variables x and y.

In what follows, we first discuss linear models and how to solve them, and then we use that knowledge to study nonlinear models.

5.6 n-DIMENSIONAL LINEAR DIFFERENCE EQUATIONS

Consider the n-dimensional linear discrete-time system

$$\begin{cases} x_1\,(t+1) = a_{11}x_1\,(t) + a_{12}x_2\,(t) + \cdots + a_{1n}x_n\,(t) \\ x_2\,(t+1) = a_{21}x_1\,(t) + a_{22}x_2\,(t) + \cdots + a_{2n}x_n\,(t) \\ \qquad\qquad\qquad\vdots \\ x_n\,(t+1) = a_{n1}x_1\,(t) + a_{n2}x_2\,(t) + \cdots + a_{nn}x_n\,(t) \end{cases}.$$

This system is called *linear* because the equations are linear in the state variables $x_1, x_2, \ldots, x_n$. It can be written as a matrix equation:

$$\begin{pmatrix} x_1\,(t+1) \\ x_2\,(t+1) \\ \vdots \\ x_n\,(t+1) \end{pmatrix} = \begin{pmatrix} a_{11} & a_{12} & \cdots & a_{1n} \\ a_{21} & a_{22} & \cdots & a_{2n} \\ \vdots & & \ddots & \vdots \\ a_{n1} & a_{n2} & \cdots & a_{nn} \end{pmatrix} \begin{pmatrix} x_1\,(t) \\ x_2\,(t) \\ \vdots \\ x_n\,(t) \end{pmatrix}.$$

The matrix equation can be written with the simpler notation

$$\mathbf{x}\,(t+1) = \mathbf{M}\mathbf{x}\,(t),$$

where

$$\mathbf{x} = \begin{pmatrix} x_1 \\ x_2 \\ \vdots \\ x_n \end{pmatrix} \text{ and } \mathbf{M} = \begin{pmatrix} a_{11} & a_{12} & \cdots & a_{1n} \\ a_{21} & a_{22} & \cdots & a_{2n} \\ \vdots & & \ddots & \vdots \\ a_{n1} & a_{n2} & \cdots & a_{nn} \end{pmatrix}.$$

5.6.1 n-Dimensional Leslie Models

Suppose a species has n life stages or "classes" of equal duration, which we choose as the model time step. Suppose that in one time step, individuals survive to move from class i to class $i+1$ with probability p_i for $i = 1, 2, \ldots, n - 1$, and that individuals in class n (the final class) survive each time step to remain in class n with probability p_n. Suppose furthermore that each class i has a per capita fecundity of f_i. We can conceptualize a population of this species with the *Leslie diagram* shown in Figure 5.2.

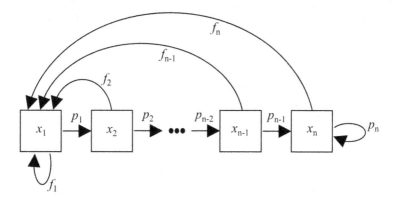

Figure 5.2: n-dimensional Leslie diagram.

If $x_i(t)$ is the number of organisms in class i at time t, we can write

$$\begin{cases} x_1(t+1) = f_1 x_1(t) + f_2 x_2(t) + \cdots + f_n x_n(t) \\ x_2(t+1) = p_1 x_1(t) \\ x_3(t+1) = p_2 x_2(t) \\ \vdots \\ x_n(t+1) = p_{n-1} x_{n-1}(t) + p_n x_n(t), \end{cases}$$

or, in matrix form,

$$\begin{pmatrix} x_1(t+1) \\ x_2(t+1) \\ \vdots \\ x_n(t+1) \end{pmatrix} = \begin{pmatrix} f_1 & f_2 & \cdots & f_n \\ p_1 & 0 & \cdots & 0 \\ 0 & p_2 & \ddots & \vdots \\ \vdots & 0 & \ddots & 0 \\ 0 & 0 & p_{n-1} & p_n \end{pmatrix} \begin{pmatrix} x_1(t) \\ x_2(t) \\ \vdots \\ x_n(t) \end{pmatrix}.$$

The matrix

$$\mathbf{M} = \begin{pmatrix} f_1 & f_2 & \cdots & f_n \\ p_1 & 0 & \cdots & 0 \\ 0 & p_2 & \ddots & \vdots \\ \vdots & 0 & \ddots & 0 \\ 0 & 0 & p_{n-1} & p_n \end{pmatrix}$$

is called a *Leslie matrix*. The birth rates are in the first row, while the survivorships are on the subdiagonal.

5.7   SOLVING LINEAR SYSTEMS OF DIFFERENCE EQUATIONS

5.7.1   An Example

We begin with an example. Consider the two-dimensional linear system

$$\begin{cases} x_{t+1} = 3x_t + 2y_t \\ y_{t+1} = \frac{1}{2}x_t + 3y_t \\ x_0 = 0 \\ y_0 = 2 \end{cases} \tag{5.3}$$

What are the equilibria of this system? The *fixed point equations*, or *equilibrium equations*, are

$$\begin{cases} x = 3x + 2y \\ y = \frac{1}{2}x + 3y \end{cases}.$$

The first of these implies that $y = -x$. The second, however, implies that $y = -x/4$. This is a contradiction unless $x = y = 0$. Thus, the only equilibrium is the extinction state $(x_e, y_e) = (0,0)$. Because the given initial condition $(x_0, y_0) = (0, 2)$ is different from the equilibrium, we know that the system is not initially at equilibrium, and so the population size must change over time. How will it change? Will it tend toward the extinction equilibrium? In order to answer this question, we will find the closed-form solution of system (5.3).

Recall that the closed-form solution to the one-dimensional linear difference equation

$$x_{t+1} = bx_t$$

is

$$x_t = cb^t,$$

where c is the initial condition. We might expect that the solution of system (5.3) will be similar. Therefore, let's look for nontrivial solutions (solutions that are not constantly zero) of the form

$$x_t = v_1\lambda^t \tag{5.4}$$
$$y_t = v_2\lambda^t,$$

where v_1, v_2, and λ are fixed real numbers. Let's require $\lambda \neq 0$, and also require that at least one of v_1 or v_2 be nonzero; otherwise, we will simply recover the equilibrium solution (trivial solution) $(0,0)$.

The proposed solution (5.4) is called the *Ansatz*, a German word for "approach." Because it is a noun in German, Ansatz is capitalized.

If we plug our proposed solution (5.4) into the dynamical system (5.3), we arrive at the following equations:

$$\begin{cases} v_1 \lambda^{t+1} = 3v_1 \lambda^t + 2v_2 \lambda^t \\ v_2 \lambda^{t+1} = \frac{1}{2}v_1 \lambda^t + 3v_2 \lambda^t \end{cases}.$$

Given $\lambda \neq 0$, we can divide through by λ^t to obtain

$$\begin{cases} v_1 \lambda = 3v_1 + 2v_2 \\ v_2 \lambda = \frac{1}{2}v_1 + 3v_2 \end{cases}.$$

One further manipulation gives us the algebraic system of equations

$$\begin{cases} (3 - \lambda)\, v_1 + 2v_2 = 0 \\ \frac{1}{2}v_1 + (3 - \lambda)\, v_2 = 0 \end{cases}. \tag{5.5}$$

We are looking for a solution (v_1, v_2) of system (5.5), where v_1, v_2 are not both zero. Notice, however, that $(v_1, v_2) = (0, 0)$ is indeed a solution of system (5.5). If you think of the two equations $(3 - \lambda)\, v_1 + 2v_2 = 0$ and $\frac{1}{2}v_1 + (3 - \lambda)\, v_2 = 0$ as lines in the v_1-v_2 plane, this means the two lines meet at the origin $(0, 0)$. But system (5.5) has a nontrivial solution (v_1, v_2) if and only if the lines meet somewhere other than the origin. This can happen if and only if the two lines are coincident. The two lines are coincident if and only if the two equations $(3 - \lambda)\, v_1 + 2v_2 = 0$ and $\frac{1}{2}v_1 + (3 - \lambda)\, v_2 = 0$ are linearly dependent, that is, if and only if they are multiples of each other. A fact from linear algebra says this is true if and only if the determinant of the coefficient matrix is zero, that is, if and only if

$$\begin{vmatrix} (3 - \lambda) & 2 \\ \frac{1}{2} & (3 - \lambda) \end{vmatrix} = 0.$$

This holds if and only if

$$\lambda^2 - 6\lambda + 8 = 0,$$

which is to say

$$(\lambda - 4)\,(\lambda - 2) = 0.$$

The last equality holds if and only if $\lambda = 2$ or $\lambda = 4$. These values of λ are called the *eigenvalues* for the system. We have shown that system (5.5) has a nontrivial solution (v_1, v_2) if and only if $\lambda = 2$ or $\lambda = 4$.

We now consider each of these eigenvalues, in turn, and find v_1 and v_2 for each. If $\lambda = 2$, then system (5.5) becomes

$$\left\{ \begin{array}{l} v_1 + 2v_2 = 0 \\ \frac{1}{2}v_1 + v_2 = 0 \end{array} \right. .$$

Note that these last two equations are indeed multiples of each other; their graphs in the v_1-v_2 plane are coincident lines. Thus, there are infinitely many solutions (v_1, v_2). However, these solutions are constrained by the relationship

$$v_1 = -2v_2.$$

For example, if $v_2 = 1$, then $v_1 = -2$. We say an *eigenvector associated with eigenvalue* $\lambda = 2$ is

$$\left(\begin{array}{c} v_1 \\ v_2 \end{array} \right) = \left(\begin{array}{c} -2 \\ 1 \end{array} \right).$$

Note that any scalar multiple of this eigenvector is also an eigenvector. Looking back at our proposed solution (5.4), we see that we have found an *eigensolution*

$$\begin{array}{rcl} x_t & = & v_1 \lambda^t = -2 \, (2)^t \\ y_t & = & v_2 \lambda^t = 1 \, (2)^t , \end{array}$$

which can also be written

$$\left(\begin{array}{c} x_t \\ y_t \end{array} \right) = \left(\begin{array}{c} -2 \\ 1 \end{array} \right) 2^t ,$$

for system (5.3).

Now let's consider the other eigenvalue, $\lambda = 4$. If $\lambda = 4$, system (5.5) becomes

$$\left\{ \begin{array}{l} -v_1 + 2v_2 = 0 \\ \frac{1}{2}v_1 - v_2 = 0 \end{array} \right. .$$

Thus, $v_1 = 2v_2$, so we can let $v_2 = 1$ and $v_1 = 2$. Therefore, an eigenvector associated with $\lambda = 4$ is

$$\begin{pmatrix} v_1 \\ v_2 \end{pmatrix} = \begin{pmatrix} 2 \\ 1 \end{pmatrix},$$

and the corresponding eigensolution for system (5.3) is

$$\begin{aligned} x_t &= 2\,(4)^t \\ y_t &= 1\,(4)^t, \end{aligned}$$

that is

$$\begin{pmatrix} x_t \\ y_t \end{pmatrix} = \begin{pmatrix} 2 \\ 1 \end{pmatrix} 4^t.$$

A fact from dynamical systems theory says that the eigensolutions *span* the space of all solutions; that is, any solution of the original system (5.3) is a linear combination of the eigensolutions. The *general solution* (i.e., the set of all solutions) of system (5.3) is therefore the closed-form solution

$$\begin{aligned} \begin{pmatrix} x_t \\ y_t \end{pmatrix} &= c_1 \begin{pmatrix} -2 \\ 1 \end{pmatrix} 2^t + c_2 \begin{pmatrix} 2 \\ 1 \end{pmatrix} 4^t \\ &= \begin{pmatrix} -2c_1\,(2)^t + 2c_2\,(4)^t \\ c_1\,(2)^t + c_2\,(4)^t \end{pmatrix}, \end{aligned}$$

where $c_1, c_2 \in R$. In component form, the general solution is

$$\begin{aligned} x_t &= -2c_1\,(2)^t + 2c_2\,(4)^t \\ y_t &= c_1\,(2)^t + c_2\,(4)^t. \end{aligned} \qquad (5.6)$$

Now we want to find the *particular solution* corresponding to the given initial values of $x_0 = 0$ and $y_0 = 2$. We set $t = 0$ in the general solution (5.6) to obtain the algebraic system

$$\begin{aligned} 0 &= -2c_1 + 2c_2 \\ 2 &= c_1 + c_2. \end{aligned}$$

From this, we see that $c_1 = 1$ and $c_2 = 1$. Plugging these values for c back into the general solution (5.6) gives the particular solution

$$\begin{aligned} x_t &= -2\,(2)^t + 2\,(4)^t \\ y_t &= 2^t + 4^t. \end{aligned}$$

This is the closed-form solution of the initial value problem (5.3). Note that the dominant eigenvalue ($\lambda = 4$) determines the exponential rate of growth for the system. In particular, the equilibrium $(0, 0)$ is unstable, because solutions starting near $(0, 0)$ grow exponentially.

5.7.2 Solving the General Two-Dimensional System

We now solve the general two-dimensional linear system

$$\begin{cases} x_{t+1} = ax_t + by_t \\ y_{t+1} = cx_t + dy_t \end{cases}. \tag{5.7}$$

Clearly, this system has the extinction state $(x, y) = (0, 0)$ as a solution. In fact, the only equilibrium is $(0, 0)$ as long as the determinant of the coefficient matrix is nonzero, that is, as long as

$$\begin{vmatrix} a & b \\ c & d \end{vmatrix} \neq 0.$$

In what follows, we will always assume this holds.

We want to look for nontrivial solutions of the form

$$\begin{aligned} x_t &= v_1 \lambda^t \\ y_t &= v_2 \lambda^t, \end{aligned} \tag{5.8}$$

with $\lambda \neq 0$ and v_1, v_2 not both zero. This is the Ansatz.

Plugging the Ansatz (5.8) into system (5.7) yields

$$\begin{cases} v_1 \lambda^{t+1} = av_1 \lambda^t + bv_2 \lambda^t \\ v_2 \lambda^{t+1} = cv_1 \lambda^t + dv_2 \lambda^t \end{cases},$$

which can be simplified to

$$\begin{cases} v_1 \lambda = av_1 + bv_2 \\ v_2 \lambda = cv_1 + dv_2 \end{cases}.$$

Rearranging, we obtain the algebraic system of equations

$$\begin{cases} (a - \lambda) v_1 + bv_2 = 0 \\ cv_1 + (d - \lambda) v_2 = 0 \end{cases}. \tag{5.9}$$

System (5.9) has the trivial solution $(v_1, v_2) = (0,0)$. It has a nontrivial solution (v_1, v_2) if and only if the two equations in (5.9) are coincident lines, if and only if

$$\begin{vmatrix} (a - \lambda) & b \\ c & (d - \lambda) \end{vmatrix} = 0,$$

that is, if and only if

$$\lambda^2 - (a + d)\lambda + (ad - bc) = 0.$$

This last equation is called the *characteristic equation*. The solutions λ_1 and λ_2 of the characteristic equation are the *eigenvalues* for the system. These are the values of λ for which the Ansatz (5.8) is a solution; hence, these are the exponential growth rates of the system. For each eigenvalue, we can find an associated *eigenvector* from system (5.9) by plugging in the eigenvalue and finding a relationship between v_1 and v_2. If $\lambda_1 \neq \lambda_2$ and $\mathbf{v} = (v_1, v_2)^\top$ and $\mathbf{w} = (w_1, w_2)^\top$ are linearly independent eigenvectors associated with λ_1 and λ_2, respectively, then the *general solution* of system (5.7) written in vector form is

$$\mathbf{x}_t = c_1 \mathbf{v} \lambda_1^t + c_2 \mathbf{w} \lambda_2^t,$$

where $\mathbf{x}_t = (x_t, y_t)^\top$ and $c_1, c_2 \in R$. In component form, the general solution is

$$\begin{aligned} x_t &= c_1 v_1 \lambda_1^t + c_2 w_1 \lambda_2^t \\ y_t &= c_1 v_2 \lambda_1^t + c_2 w_2 \lambda_2^t. \end{aligned}$$

A given particular solution is found by plugging the initial conditions into the general solution, solving for the constants c_1, c_2, and plugging these back into the general solution.

5.7.3 Solving Higher-Dimensional Systems

Linear systems with dimension higher than two are solved in the same way as two-dimensional systems. A discrete linear model

$$\begin{cases} x_1(t+1) = a_{11}x_1(t) + a_{12}x_2(t) + \cdots + a_{1n}x_n(t) \\ x_2(t+1) = a_{21}x_1(t) + a_{22}x_2(t) + \cdots + a_{2n}x_n(t) \\ \quad \vdots \\ x_n(t+1) = a_{n1}x_1(t) + a_{n2}x_2(t) + \cdots + a_{nn}x_n(t) \end{cases} \quad (5.10)$$

can be written as a matrix equation

$$\mathbf{x}\,(t+1) = \mathbf{M}\mathbf{x}\,(t)\,,$$

where

$$\mathbf{x} = \begin{pmatrix} x_1 \\ x_2 \\ \vdots \\ x_n \end{pmatrix} \quad \text{and } \mathbf{M} = \begin{pmatrix} a_{11} & a_{12} & \cdots & a_{1n} \\ a_{21} & a_{22} & \cdots & a_{2n} \\ \vdots & & \ddots & \vdots \\ a_{n1} & a_{n2} & \cdots & a_{nn} \end{pmatrix}.$$

The *eigenvalues* of the $n \times n$ matrix $\mathbf{M}$ are the solutions λ of the *characteristic equation*

$$|\mathbf{M} - \lambda\mathbf{I}| = 0.$$

If you do not know how to find the determinant of a $n \times n$ matrix for $n > 2$, you can look this up in an introductory linear algebra text or calculus text.

Given a particular eigenvalue λ, a solution $\mathbf{v}$ of the vector equation

$$(\mathbf{M} - \lambda\mathbf{I})\,\mathbf{v} = \mathbf{0}$$

is called an *eigenvector corresponding to* λ.

Let $\lambda_1, \lambda_2, \ldots, \lambda_n$ be distinct eigenvalues of $\mathbf{M}$ with corresponding eigenvectors $\mathbf{v}_1, \mathbf{v}_2, \cdots, \mathbf{v}_n$. Suppose the eigenvectors are linearly independent (i.e., no one of them can be written as a linear combination of the others); this is true in many applications. Then the *general solution* of system (5.10) can be written in vector form as

$$\mathbf{x}\,(t) = c_1\mathbf{v}_1\lambda_1^t + c_2\mathbf{v}_2\lambda_2^t + \cdots + c_n\mathbf{v}_n\lambda_n^t. \tag{5.11}$$

Initial conditions give rise to particular values of the constants c_i, leading to the *particular solution* of the initial value problem.

The *fundamental theorem of demography* basically says that the magnitudes of the eigenvalues tell you whether or not the population is going extinct. The theorem has two parts:

1. If all the eigenvalues are inside the unit circle in the complex plane, that is, if $|\lambda_i| < 1$ for all $i = 1, 2, \cdots n$, then the extinction equilibrium is asymptotically stable and all solutions approach the zero vector.

2. If any one of the eigenvalues is outside the unit circle, that is, if $|\lambda_i| > 1$ for some i, then the extinction equilibrium is unstable.

The *strong ergodic property* tells us how the age classes are distributed in the long run: If there is a *simple dominant eigenvalue* λ_1, that is, an unrepeated eigenvalue λ_1 such that $|\lambda_1| > |\lambda_i|$ for every $i \neq 1$, then from equation (5.11) we have (Exercise 4)

$$\lim_{t \to \infty} \frac{1}{\lambda_1^t} \mathbf{x}(t) = c_1 \mathbf{v}_1.$$

The eigenvector $\mathbf{v}_1$ belonging to the simple dominant eigenvalue λ_1 gives the *stable age distribution* (or stable stage distribution) of the model, that is, the asymptotic relative distribution of stages as $t \to \infty$. *This does not necessarily mean the population is equilibrating*; it can be growing exponentially or declining to extinction, depending on the magnitude of λ_1. For example, if the dominant eigenvalue is $\lambda_1 = 1.5$ and its eigenvector is $(2, 1)^\mathsf{T}$, then in the long run, the population is growing without bound with an asymptotic age distribution of twice as many juveniles as adults.

5.8 NONLINEAR SYSTEMS

Let's consider a nonlinear Leslie model for populations of the flour beetle *Tribolium castaneum*. Under laboratory conditions, eggs quickly hatch into larvae, larvae pupate after about 14 days, and pupae metamorphize into adults after another 14 days. Larvae and adults eat eggs, and adults eat pupae. A well-validated model of *Tribolium castaneum* dynamics is the three-dimensional nonlinear Leslie model known as the *LPA model* (Cushing et al. 2003):

$$
\begin{aligned}
L_{t+1} &= bA_t e^{-c_{el}L_t - c_{ea}A_t} \\
P_{t+1} &= (1 - \mu_l) L_t \\
A_{t+1} &= P_t e^{-c_{pa}A_t} + (1 - \mu_a) A_t.
\end{aligned}
$$

The LPA model is nonlinear because of the Ricker nonlinearities in the first and third equations.

Nonlinear models such as the LPA model do not, in general, have simple closed-form solutions. Furthermore, it is often impossible to solve explicitly for the equilibria; in such cases, the equilibria must be found numerically. Given an equilibrium, however, we can study its stability through the technique of linearization, just as we did for one-dimensional maps.

5.8.1 Linearization

Consider the two-dimensional nonlinear system

$$
\begin{aligned}
x_{t+1} &= f(x_t, y_t) \\
y_{t+1} &= g(x_t, y_t).
\end{aligned}
\tag{5.12}
$$

Equilibria of this system are constant solutions, i.e., constant pairs (x_e, y_e) that simultaneously satisfy the fixed point system of equations

$$
\begin{aligned}
x_e &= f(x_e, y_e) \\
y_e &= g(x_e, y_e).
\end{aligned}
$$

We want to approximate the behavior of the system near the equilibria in order to investigate the stability of the equilibria.

In multivariate calculus, one learns to approximate a surface $z = f(x, y)$ near the point $(x, y) = (a, b)$ with a plane that is tangent to the surface at that point. The equation of the tangent plane is

$$
z = f(a, b) + \frac{\partial f}{\partial x}(a, b)(x - a) + \frac{\partial f}{\partial y}(a, b)(y - b),
$$

and this is a good approximation for $f(x, y)$ as long as the point (x, y) is sufficiently close to the point (a, b). Here, the notation $\partial f / \partial x$ refers to the partial derivative of f with respect to the variable x. Computationally, $\partial f / \partial x$ is simply the derivative of $f(x, y)$ with respect to x while treating y as a constant. To practice computing partial derivatives, see any multivariate calculus text.

If we approximate f and g with their tangent planes at equilibrium (x_e, y_e), we have

$$
\begin{aligned}
f(x,y) &\approx f(x_e, y_e) + \frac{\partial f}{\partial x}(x_e, y_e)(x - x_e) + \frac{\partial f}{\partial y}(x_e, y_e)(y - y_e) \\
&= x_e + \frac{\partial f}{\partial x}(x_e, y_e)(x - x_e) + \frac{\partial f}{\partial y}(x_e, y_e)(y - y_e) \\
g(x,y) &\approx g(x_e, y_e) + \frac{\partial g}{\partial x}(x_e, y_e)(x - x_e) + \frac{\partial g}{\partial y}(x_e, y_e)(y - y_e) \\
&= y_e + \frac{\partial g}{\partial x}(x_e, y_e)(x - x_e) + \frac{\partial g}{\partial y}(x_e, y_e)(y - y_e)
\end{aligned}
$$

for $x \approx x_e$ and $y \approx y_e$. Thus, system (5.12) can be approximated by

$$
\begin{aligned}
x_{t+1} &\approx x_e + \frac{\partial f}{\partial x}(x_e, y_e)(x_t - x_e) + \frac{\partial f}{\partial y}(x_e, y_e)(y_t - y_e) \\
y_{t+1} &\approx y_e + \frac{\partial g}{\partial x}(x_e, y_e)(x_t - x_e) + \frac{\partial g}{\partial y}(x_e, y_e)(y_t - y_e)
\end{aligned}
$$

for $x_t \approx x_e$ and $y_t \approx y_e$.

The change of variables

$$
\begin{aligned}
u_t &= x_t - x_e \\
v_t &= y_t - y_e
\end{aligned}
\tag{5.13}
$$

leads to the linear system

$$
\begin{aligned}
u_{t+1} &\approx \frac{\partial f}{\partial x}(x_e, y_e)\, u_t + \frac{\partial f}{\partial y}(x_e, y_e)\, v_t \\
v_{t+1} &\approx \frac{\partial g}{\partial x}(x_e, y_e)\, u_t + \frac{\partial g}{\partial y}(x_e, y_e)\, v_t
\end{aligned}
$$

for $u_t \approx 0$ and $v_t \approx 0$. Note that, by their definition in equation (5.13), the variables u_t and v_t measure the displacement, or variation, of the system from equilibrium.

Definition 5.1 *The **linearization** of a nonlinear system*

$$
\begin{aligned}
x_{t+1} &= f(x_t, y_t) \\
y_{t+1} &= g(x_t, y_t)
\end{aligned}
$$

at an equilibrium (x_e, y_e) *is the linear system*

$$u_{t+1} = \frac{\partial f}{\partial x}(x_e, y_e) u_t + \frac{\partial f}{\partial y}(x_e, y_e) v_t$$

$$v_{t+1} = \frac{\partial g}{\partial x}(x_e, y_e) u_t + \frac{\partial g}{\partial y}(x_e, y_e) v_t.$$

We can write the linearization as

$$\mathbf{u}_{t+1} = \mathbf{J}(x_e, y_e)\mathbf{u}_t,$$

where

$$\mathbf{u} = \begin{pmatrix} u \\ v \end{pmatrix} \text{ and } \mathbf{J} = \begin{pmatrix} \frac{\partial f}{\partial x} & \frac{\partial f}{\partial y} \\ \frac{\partial g}{\partial x} & \frac{\partial g}{\partial y} \end{pmatrix}.$$

The matrix $\mathbf{J}$ is called the *Jacobian matrix*.

5.8.2 An Example

Let's carry out a complete equilibrium stability analysis for the nonlinear system

$$x_{t+1} = y_t \tag{5.14}$$

$$y_{t+1} = \frac{3}{2}y_t(1 - x_t)$$

at each of its equilibria.

First, we find the equilibria by solving the fixed point equation

$$x = y$$

$$y = \frac{3}{2}y(1 - x).$$

The equilibria are $(0,0)$ and $(1/3, 1/3)$. Then we compute the Jacobian:

$$\mathbf{J} = \begin{pmatrix} \frac{\partial f}{\partial x} & \frac{\partial f}{\partial y} \\ \frac{\partial g}{\partial x} & \frac{\partial g}{\partial y} \end{pmatrix}$$

$$= \begin{pmatrix} 0 & 1 \\ -\frac{3}{2}y & \frac{3}{2}(1-x) \end{pmatrix}.$$

At the equilibrium $(0,0)$, the Jacobian is

$$\mathbf{J}(0,0) = \begin{pmatrix} 0 & 1 \\ 0 & \frac{3}{2} \end{pmatrix},$$

so the linearization at $(0,0)$ is the linear system

$$
\begin{aligned}
u_{t+1} &= v_t \\
v_{t+1} &= \frac{3}{2}v_t.
\end{aligned}
$$

The eigenvalues of the linearized system are $\lambda = 0, 3/2$. Since $\lambda = 3/2$ is outside the unit circle in the complex plane (that is, $|\lambda| > 1$), the $(0,0)$ equilibrium of the linearized system is unstable.

At the equilibrium $(1/3, 1/3)$, the Jacobian is

$$
\mathbf{J} = \begin{pmatrix} 0 & 1 \\ -\frac{1}{2} & 1 \end{pmatrix},
$$

so the linearization at $(1/3, 1/3)$ is

$$
\begin{aligned}
u_{t+1} &= v_t \\
v_{t+1} &= -\frac{1}{2}u_t + v_t.
\end{aligned}
$$

The eigenvalues of this linear system are $\lambda = \frac{1}{2} \pm \frac{1}{2}i$, which are both inside the unit circle in the complex plane since $|\lambda| = \sqrt{\left(\frac{1}{2}\right)^2 + \left(\frac{1}{2}\right)^2} = \frac{\sqrt{2}}{2} < 1$. Thus, the $(0,0)$ equilibrium of the linearized u-v system is stable. Since u_t and v_t measure the displacement of the system away from the equilibrium $(1/3, 1/3)$, it seems reasonable to suspect that the $(1/3, 1/3)$ equilibrium of the nonlinear system (5.14) is stable as well. The following theorem states that this is so.

Definition 5.2 *An equilibrium (x_e, y_e) of*

$$
\begin{aligned}
x_{t+1} &= f(x_t, y_t) \\
y_{t+1} &= g(x_t, y_t)
\end{aligned}
$$

*is **hyperbolic** if and only if the Jacobian matrix $\mathbf{J}(x_e, y_e)$ evaluated at the fixed point has no eigenvalues of modulus one.*

Theorem 5.1 *(Linearization Theorem) Let $f(x, y)$ and $g(x, y)$ be continuously differentiable functions of x and y (that is, have* *continuous derivatives in both variables). Consider the nonlinear system*

$$x_{t+1} = f(x_t, y_t)$$
$$y_{t+1} = g(x_t, y_t)$$

with hyperbolic fixed point (x_e, y_e).

1. *If all the eigenvalues of the Jacobian matrix $\mathbf{J}(x_e, y_e)$ at the fixed point have moduli less than one, then the fixed point (x_e, y_e) is asymptotically stable.*

2. *If at least one of the eigenvalues of $\mathbf{J}(x_e, y_e)$ has modulus greater than one, then (x_e, y_e) is unstable.*

5.9   EXERCISES

1. Consider the discrete-time system

$$x_{t+1} = 2x_t + 3y_t$$
$$y_{t+1} = -x_t - 2y_t.$$

a. Find the general solution.

b. Find the particular solution subject to the initial conditions $x_0 = 4$ and $y_0 = 0$.

2. Consider the discrete-time system

$$x_{t+1} = 2x_t + 2y_t$$
$$y_{t+1} = 7x_t - 3y_t.$$

a. Find the general solution.

b. Find the particular solution subject to the initial conditions $x_0 = 0$ and $y_0 = 9$.

3. Consider the discrete-time system

$$x_{t+1} = 18x_t - 11y_t \qquad (5.15)$$
$$y_{t+1} = 8x_t - y_t.$$

a. Write a program that *iterates* system (5.15). Use the initial conditions $x_0 = 35$ and $y_0 = 26$, and compute x_t and y_t for $t = 1$ to $t = 5$. In the program, create a 6×4 matrix with the following four columns: a column for t which would begin with 0 and end with 5; a column for the corresponding value of x_t; a column for the corresponding value of y_t; and a column for the value of the ratio x_t/y_t. Output this matrix to the screen. Attach your program and the output matrix.

b. Find the general solution of system (5.15).

c. Find the dominant eigenvalue and the stable age distribution.

d. How does the stable age distribution from part (3c) relate to the x_t/y_t column in your output from part (3a)?

e. Find the particular solution subject to the initial conditions $x_0 = 35$ and $y_0 = 26$.

f. Evaluate the particular solution in part (3e) at $t = 5$ to check the values of x_5 and y_5 that you obtained by iteration in part (3a).

g. Consider the particular solution you found in part (3e). If you wrote it in vector form, rewrite it in component form. Use the closed-form expressions for x_t and y_t to write a closed-form expression for x_t/y_t. Now use the methods of calculus to compute $\lim_{t \to \infty} x_t/y_t$. How does this limit relate to the stable age distribution you found in part (3c)?

4. Prove that if λ_1 is the dominant eigenvalue for system (5.10), then the strong ergodic property holds, that is, $\lim_{t \to \infty} \frac{1}{\lambda_1^t} \mathbf{x}(t) = c_1 \mathbf{v}_1$.

5. Consider a population categorized into three age classes: juveniles (ages zero to one year), one-year-olds (ages one to two years), and two-year-olds (ages two to three years). Suppose individuals mature sexually at one year and live no longer than three years. Suppose that the average per capita number of births per year for one-year-olds is 4 and the average per capita number of births per year for two-year-olds is 2/3. Suppose also that 60% of juveniles do not survive until sexual maturity, and that 15% of one-year-olds do not survive to become two-year-olds.

    a. Draw and label a Leslie diagram for this population.

    b. Give the Leslie matrix model for this population.

    c. Suppose the population starts with 10 two-year-olds and no individuals of any other age. *Iterate* the Leslie model to find the number of animals in each age class and the total population size after 3 years.

    d. Find the long-term growth rate (dominant eigenvalue) and the stable age distribution (corresponding eigenvector).

    e. Find the general solution for the Leslie model.

    f. Find the particular solution, given the initial condition in (5c).

    g. What is the long-term fate of the population?

6. Consider a stage-structured juvenile-adult system in which the time step is chosen to be the duration of the juvenile stage. Assume that after one time step, 75% of the juveniles survive to become sexually mature adults. Suppose that the average per capita birth rate for adults is $b = 1/6$, and that 25% of adults survive each time step.

    a. Draw and label a Leslie diagram for this system.

    b. Give the Leslie matrix model for this system.

    c. Find the general solution.

    d. What is the long-term fate of this population, according to the Leslie model prediction?

    e. Suppose we begin with a juvenile density of 3, but with 0 adults. Find the particular solution for the system. Use a computer to *iterate* the Leslie model to $t = 20$. Graph the numbers of juveniles and adults in the population as a function of time. Attach your graph.

    f. What is the stable age distribution for this population?

7. In this problem, you will analyze a juvenile-adult model for a cannibalistic insect population. The model is a discrete-time model, with a time step of one week. The larvae (juveniles) of

this insect damage agricultural crops. The model with initial conditions is

$$
\begin{aligned}
J_{t+1} &= bA_t \exp\left(-c_{ej}J_t - c_{ea}A_t\right) \\
A_{t+1} &= J_t \exp\left(-c_{ja}A_t\right) + (1 - \mu_a)A_t \qquad (5.16) \\
J_0 &= 500 \text{ juveniles per unit area} \\
A_0 &= 50 \text{ adults per unit area}
\end{aligned}
$$

a. Write down a list of assumptions on which model (5.16) is based.

b. Draw and label a Leslie diagram for model (5.16).

c. List the parameters in model (5.16).

d. Write down the fixed point equations for model (5.16).

e. From your answer in (7d), prove that the extinction state $(J, A) = (0, 0)$ is an equilibrium.

f. Find the linearization of model (5.16) at the extinction state $(0, 0)$ by hand. Hint: Your answer must be a linear system of difference equations. It will have parameters in it.

g. Find conditions on the parameters that guarantee the stability/instability of the extinction state.

h. Define the *inherent net reproductive number* n to be $n = b/\mu_a$. Because the expected reproductive lifespan of an adult is $1/\mu_a$, we can interpret n to be *the expected number of offspring of an adult over its lifetime*. Rewrite your stability conditions in (7g) in terms of the inherent net reproductive number n, and give a biological interpretation of these conditions.

i. Write a computer program to simulate (iterate) model (5.16). Attach your program.

j. Suppose you gather data to parameterize the model, and estimate that $b = 30$, $c_{ej} = c_{ea} = c_{ja} = 0.01$, and $\mu_a = 0.02$. Use your program to graph the predicted time series for the number of juveniles over the first 30 weeks. Attach your graph. What is the long-term fate of the population? Is this consistent with your answer in (7g)?

k. In an effort to reduce crop damage, an agricultural agent applies an insecticide that kills adult insects. Suppose this raises the adult death rate to $\mu_a = 0.20$ while leaving the other parameters unchanged. Re-graph the time series. What change occurs in the population dynamics? What changes in the insect infestation does the farmer observe? How will these changes affect crop damage?

l. The agricultural agent now applies a higher level of insecticide so that $\mu_a = 0.98$. (The adult death rate is 98%!) Graph the time series again. What change occurs in the population dynamics? What changes in the insect infestation does the farmer observe?

m. Use the computer to draw a bifurcation diagram for larval numbers, using μ_a as a bifurcation parameter with the range $0 \le \mu_a \le 1$. Use this diagram to explain why an insecticide that affects only the adult death rate will probably not reduce crop damage.

n. The agricultural agent now decides to use a different kind of insecticide—one that reduces the birth rate b. Given that the natural value of μ_a is 0.02, how much must the scientist reduce the value of b to cause extinction of the insect pest? Hint: Use your answer in (7g).

o. Explain mathematically, from inspecting model (5.16), why reducing the number of adults can actually increase crop damage.

p. Explain biologically why reducing the number of adults can actually increase crop damage.

q. How did this exercise affect your thinking about population dynamics and mathematical modeling?

r. Consider once again your fixed point equations in (7d) with arbitrary parameter values. Prove that if $b > \mu_a$, then there exists a unique nonzero fixed point. Hint: Two expressions are equal where their curves intersect.

s. Let $b = 30$, $c_{ej} = c_{ea} = c_{ja} = 0.01$, and $\mu_a = 0.02$. Numerically estimate the value of the nonzero fixed point. Attach your program.

t. Find the linearization of the system at the nonzero fixed point in (7s). Your answer must be a linear system of difference equations.

u. Is this nonzero equilibrium stable or unstable?

BIBLIOGRAPHY

Caswell, H. 2001. *Matrix Population Models: Construction, Analysis, and Interpretation*, 2nd ed. Sinauer Associates, Sunderland, MA. [Caswell's classic book should be in the library of every ecologist and mathematical biologist.]

Cushing, J. M., Costantino, R. F., Dennis, B., Desharnais, R. A., and Henson, S. M. 2003. *Chaos in Ecology: Experimental Nonlinear Dynamics.* Academic Press, San Diego, CA. [Summarizes the large body of work involved in the documentation of chaos in a laboratory population. In Chapter 6 we will draw heavily on this book to showcase how dynamic population data can be connected to mathematical models.]

CHAPTER 6

Flour Beetle Dynamics: A Case Study

6.1 WHAT YOU SHOULD KNOW ABOUT THIS CHAPTER

This chapter is based on a body of research conducted by a collaboration of scientists and mathematicians known as the "Beetle Team." The Beetle Team used laboratory populations of flour beetles (*Tribolium castaneum*) to test nonlinear dynamics theory. "The Beetles" initially were composed of ecologist R. F. (Bob) Costantino of the University of Rhode Island; mathematician J. M. (Jim) Cushing of the University of Arizona; statistical ecologist Brian Dennis of the University of Idaho; and ecologist R. A. (Bob) Desharnais of California State University, Los Angeles. Later, mathematicians Shandelle M. Henson and Aaron A. King, then postdocs at the University of Arizona, joined the Team. Key research papers authored by the group, including those on which this chapter is based, are found in this chapter's bibliography.

The ideas, techniques, and coding in this case study are similar to those in the avian bone growth case study in Chapter 2, except that here you will be applying the ideas to a discrete-time dynamical system with three state variables. You will need the skills from Chapter 2 and those from all three appendices.

6.2 FLOUR BEETLES

Flour beetles of the genus *Tribolium* are small insects of the family Tenebrionidae. They can be found occasionally in their native forest

DOI: 10.1201/9781003265382-8

Figure 6.1: *T. castaneum* pupa and adult. (Used with permission of R. F. Costantino.)

habitats, but for thousands of years their preferred habitat has been milled flour and other stored grain products. These insects are hardy and prolific animals, traits which make them important agricultural pests and quite resistant to eradication. These same traits, however, make them easy to culture and manipulate in the laboratory. Flour beetles have been used extensively as animal models in genetics and population ecology. The details of their life cycle are well known. A comprehensive reference for the life history of flour beetles is Alexander Sokoloff's three-volume book "The Biology of *Tribolium*" (Sokoloff 1972; Sokoloff 1975; Sokoloff 1978). A fairly comprehensive history of the mathematics of *Tribolium* prior to the work presented in this chapter can be found in Costantino and Desharnais (1991).

Flour beetles undergo *complete metamorphosis*, which encompasses four life-cycle stages: egg, larva, pupa, and adult. The tiny eggs—as many as 500 from a single female—hatch in approximately four days under laboratory conditions and can barely be seen by the unaided eye. The larvae feed voraciously and eventually grow to the size of rice grains. After about 14 days, a larva encloses itself in a pupal case, which consists of a thin outer covering, the pupal cuticle. During the pupal stage, the individual completely reorganizes itself to become an adult (Figure 6.1). The new adult emerges another 14 days later and sheds its pupal case as little pieces of dried tissue called frass.

One might expect that populations of flour beetles, when cultured under constant environmental conditions, would grow exponentially at first and then equilibrate due to resource limitations. But they do not. Instead, their numbers typically exhibit sustained oscillations. The mechanism that regulates population size and causes these oscillations is cannibalism. In general, the mobile stages (larvae and adults) cannibalize the immobile stages (eggs and pupae). These cannibalistic encounters are crucial to the understanding of flour beetle population dynamics.

Cannibalism occurs in a broad range of taxa. At least 1,300 species, including protozoans, insects, reptiles, fish, birds, and mammals, engage in cannibalism. Some key references on cannibalism include Dong and Polis (1992); Elgar and Crespi (1992); Fox (1975); and Polis (1981).

Cannibalism often is associated with low food supply. In flour beetles, however, cannibalism does not appear to be a significant source of nutrition, nor does it appear to be caused by limitations of food or space. Instead, the incidence of this behavior is under genetic control. Varying degrees of cannibalism can be artificially selected for in the laboratory to create different genetic strains, some highly cannibalistic, and others essentially noncannibalistic. As far as the modeling process is concerned, the Beetle Team assumed *Tribolium* cannibalism occurs at random, as the mobile larvae and adults encounter immobile eggs and pupae. Under this assumption, the rate of cannibalism in *Tribolium* should be inversely proportional to habitat volume; this was established theoretically in Henson and Cushing (1997) and tested empirically using periodic habitat volumes in the laboratory (Henson et al. 1999).

6.3 DATA

The time series in Data Set 6.1 are the control replicates from laboratory experiments conducted by R. A. Desharnais in 1978 at the University of Rhode Island when he did his master of science in zoology with R. F. Costantino (Desharnais and Costantino 1980). Populations of the corn oil-sensitive strain of *Tribolium castaneum* (Herbst) were cultured in half-pint milk bottles containing 20 g of corn oil media (90% wheat flour, 5% dried brewer's yeast, and 5% liquid corn oil). The cultures were kept in an unlighted incubator at $33°C\pm1°C$ and a relative humidity of $56\% \pm 11\%$. Every two weeks, all life-cycle stages (except eggs) were censused, and the animals (including eggs) were returned to the incubator in fresh media. Small (feeding) larvae were counted as L-stage; large (nonfeeding) larvae, pupae, and callow (immature) adults were counted as P-stage; and sexually mature adults were counted as A-stage. In this chapter, we refer to L-stage animals as "larvae," P-stage animals as "pupae," and A-stage animals as "adults" for the sake of simplicity. These data are also available in Appendix A of the book by Cushing et al. (2003), along with other *Tribolium* data sets.

6.4 ASSUMPTIONS

A major goal of modeling, often called the "art of modeling," is to construct a simple, mechanistic, low-dimensional (i.e., few state variables) model that describes enough of the observed variability so that the leftover, "higher-order" variability is relatively small and can be described statistically as "noise." The beetle system has four life stages (egg, larval, pupal, and adult) and thus might require four state variables. The egg stage duration, however, is much shorter than that of the other three life stages. Following the KISS principle (keep it simple stupid), here we model the beetle system with two alternative three-dimensional models. We will select the best of these two models and attempt to validate it (Exercises 5–8).

The Beetle Team used the following assumptions.

6.4.1 Deterministic Assumptions

The deterministic assumptions were as follows:

(A1) Given that the duration of the egg stage is much shorter than the duration of the larval and pupal stages, we can ignore numbers of eggs and treat egg cannibalism as a factor that dampens recruitment into the larval stage.

(A2) In the absence of cannibalism, the per capita larval recruitment rate is a constant $b > 0$ larvae per adult per unit time.

(A3) In the presence of cannibalism, the larval recruitment rate is decreased due to the cannibalism of eggs by larvae. It is decreased by a Ricker factor of the form $e^{-c_{el}L}$ in the presence of L larvae.

(A4) In the presence of cannibalism, the larval recruitment rate is decreased due to cannibalism of eggs by adults. It is decreased by a Ricker factor of the form $e^{-c_{ea}A}$ in the presence of A adults.

(A5) A fraction μ_l of larvae die per unit time, where $0 < \mu_l < 1$.

(A6) In the absence of cannibalism, no pupae die per unit time.

(A7) In the presence of cannibalism, pupal survivorship per unit time is decreased by a fraction $e^{-c_{pa}A}$ in the presence of A adults due to the cannibalism of pupae by adults.

(A8a) There is no cannibalism of pupae by larvae.

(A8b) (Alternate hypothesis) In the presence of cannibalism, pupal survivorship per unit time is decreased by a fraction $e^{-c_{pl}L}$ in the presence of L larvae due to the cannibalism of pupae by larvae.

(A9) A fraction μ_a of adults die per unit time, where $0 < \mu_a < 1$.

6.4.2 Stochastic Assumptions

The stochastic assumptions were as follows:

(A10) There is no measurement error.

(A11a) Environmental stochasticity is the main source of noise in the system.

(A11b) (Alternate hypothesis) Demographic stochasticity is the main source of noise in the system.

(A12) Stochastic perturbations are uncorrelated in time. We also assume, for simplicity, that stochastic perturbations are uncorrelated across life-cycle stages.

Although, for the sake of simplicity, we will assume in this chapter that noise is uncorrelated across life-cycle stages, assumption (A12), it is important to note that for environmental noise, covariances between the life stages are likely. See Dennis et al. (1995) for details on estimating a covariance matrix when parameterizing the LPA model under the assumption of environmental stochasticity. Here, we assume the covariance matrix is negligible.

6.5 ALTERNATIVE DETERMINISTIC MODELS

Let

$$
\begin{aligned}
L_t &= \text{Number of larvae at time } t \\
P_t &= \text{Number of pupae at time } t \\
A_t &= \text{Number of adults at time } t.
\end{aligned}
$$

The assumptions (A1)–(A9), using (A8a), can be conceptualized in the Leslie diagram shown in Figure 6.2. The corresponding nonlinear Leslie model is known as the *LPA model*:

$$
\begin{aligned}
L_{t+1} &= bA_t e^{-c_{el} L_t - c_{ea} A_t} \\
P_{t+1} &= (1 - \mu_l) L_t \\
A_{t+1} &= P_t e^{-c_{pa} A_t} + (1 - \mu_a) A_t
\end{aligned} \tag{6.1}
$$

with time step two weeks, the duration of each of the larval and pupal stages.

The assumptions (A1)–(A9), using the alternative hypothesis (A8b), give rise to the slightly more complicated alternative model

$$
\begin{aligned}
L_{t+1} &= bA_t e^{-c_{el} L_t - c_{ea} A_t} \\
P_{t+1} &= (1 - \mu_l) L_t \\
A_{t+1} &= P_t e^{-c_{pl} L_t - c_{pa} A_t} + (1 - \mu_a) A_t,
\end{aligned} \tag{6.2}
$$

which we will call the "LPAalt model" for the purposes of this chapter.

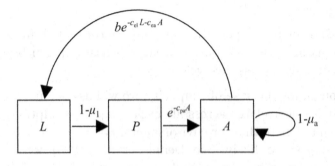

Figure 6.2: Leslie diagram for *T. castaneum*. The per capita transition rate is shown beside each arrow. The time step is 2 weeks.

6.6 STOCHASTIC MODELS

Given a discrete-time model of the form

$$x_{t+1} = f(x_t), \tag{6.3}$$

where x_t is a real number or a vector of real numbers, how does one include stochasticity? If noise is additive Gaussian, then the stochastic version of the model is

$$X_{t+1} = f(X_t) + \sigma E_t, \tag{6.4}$$

where X_t is a random variable, the function f is the deterministic predictor, $\sigma > 0$, and E_t is a standard normal random variable (mean zero and standard deviation one). The standard deviation of the noise (that is, of the random variable σE_t) is σ. When σ is set to zero, the deterministic skeleton (6.3) is recovered. Stochastic models of the form (6.4) are called *nonlinear autoregressive* (NLAR) models and have nice statistical properties.

As a matter of fact, however, process noise, whether environmental or demographic, is not in general additive. In general, data and predictions must be transformed so that the noise becomes approximately additive on some scale. A result from statistical theory, which we will simply reference here, makes it possible to approximate environmental and demographic noise with NLAR models. First, *environmental noise is approximately additive on the log scale*, that is,

$$\ln X_{t+1} = \ln(f(X_t)) + \sigma E_t,$$

whereas *demographic noise is approximately additive on the square root scale* (Cushing et al. 2003), that is,

$$\sqrt{X_{t+1}} = \sqrt{f(X_t)} + \sigma E_t.$$

In general, if a stochastic system with deterministic skeleton (6.3) can be approximated by an NLAR model of the form

$$\phi(X_{t+1}) = \phi(f(X_t)) + \sigma E_t$$

for some function ϕ, then ϕ is called a *variance-stabilizing transformation* for that system.

Under assumption (A11a), environmental stochasticity is the main source of variability in the system. The corresponding NLAR LPA model is therefore

$$
\begin{aligned}
\ln(L_{t+1}) &= \ln\left(bA_t e^{-c_{el}L_t - c_{ea}A_t}\right) + \sigma_L E_{Lt} \\
\ln(P_{t+1}) &= \ln\left((1-\mu_l)L_t\right) + \sigma_P E_{Pt} \\
\ln(A_{t+1}) &= \ln\left(P_t e^{-c_{pa}A_t} + (1-\mu_a)A_t\right) + \sigma_A E_{At},
\end{aligned}
$$

where the E_{it} are standard normal random variables (mean zero and standard deviation one). This model can be rewritten as

$$
\begin{aligned}
L_{t+1} &= bA_t e^{-c_{el}L_t - c_{ea}A_t} e^{\sigma_L E_{Lt}} \\
P_{t+1} &= (1-\mu_l)L_t e^{\sigma_P E_{Pt}} \\
A_{t+1} &= \left(P_t e^{-c_{pa}A_t} + (1-\mu_a)A_t\right) e^{\sigma_A E_{At}}.
\end{aligned} \tag{6.5}
$$

This is the "environmental noise LPA model." The "environmental noise LPAalt model" is similar (Exercise 3).

If the demographic stochasticity were the main source of variability in the system, assumption (A11b), the appropriate NLAR LPA model would be

$$
\begin{aligned}
\sqrt{L_{t+1}} &= \sqrt{bA_t e^{-c_{el}L_t - c_{ea}A_t}} + \sigma_L E_{Lt} \\
\sqrt{P_{t+1}} &= \sqrt{(1-\mu_l)L_t} + \sigma_P E_{Pt} \\
\sqrt{A_{t+1}} &= \sqrt{P_t e^{-c_{pa}A_t} + (1-\mu_a)A_t} + \sigma_A E_{At},
\end{aligned}
$$

where again the E_{it} are standard normal random variables. This model can be rewritten as

$$
\begin{aligned}
L_{t+1} &= \left(\sqrt{bA_t e^{-c_{el}L_t - c_{ea}A_t}} + \sigma_L E_{Lt}\right)^2 \\
P_{t+1} &= \left(\sqrt{(1-\mu_l)L_t} + \sigma_P E_{Pt}\right)^2 \\
A_{t+1} &= \left(\sqrt{P_t e^{-c_{pa}A_t} + (1-\mu_a)A_t} + \sigma_A E_{At}\right)^2,
\end{aligned} \tag{6.6}
$$

where the quantities inside the squares on the right-hand sides of equation (6.6) are set to zero if they are negative. This is the "demographic noise LPA model." The "demographic noise LPAalt model" is similar (Exercise 3).

6.7   MODEL PARAMETERIZATION

In order to decide whether the LPA model (6.1) or the LPAalt model
(6.2) is better, we must fit both models to the data in Data Set 6.1.
In this section, we will focus on the procedure for parameterizing the
LPA model.

There are six parameters in the deterministic LPA model. We write
them here in the order they appear in the model: $b, c_{el}, c_{ea}, \mu_l, c_{pa}, \mu_a$.
We will use the historical beetle data in Data Set 6.1 to estimate the
values of these parameters using the method of nonlinear conditioned
least squares (CLS). Ideally, we should randomly divide the data
set into two parts. One part would be used for model calibration
(parameter estimation), and the other part would be reserved for
independent model evaluation (validation). In this case, however, we
are estimating six parameters, so we need a lot of data to calibrate
the model. We will therefore use all of the historical data for model
calibration.

6.7.1   Conditioned One-Step Residuals

Let (l_t, p_t, a_t) and $(l_{t+1}, p_{t+1}, a_{t+1})$ be the actual observed data triples
at times t and $t+1$. Since we are thinking of the stochastic model as a
surrogate for the biological system, we can think of these data triples
as realizations of the random variable triple (L, P, A). The conditional
one-step residual errors at time $t+1$ are the (transformed) actual
measurements $(l_{t+1}, p_{t+1}, a_{t+1})$ at time $t+1$ minus the (transformed)
values predicted by the deterministic skeleton, given the data point
(l_t, p_t, a_t) at time t. For example, for the environmental noise LPA
model, we compute the *log residuals*

$$
\begin{aligned}
\varepsilon_{Lt} &= \ln\left(l_{t+1}\right) - \ln\left(ba_t e^{-c_{el}l_t - c_{ea}a_t}\right) \\
\varepsilon_{Pt} &= \ln\left(p_{t+1}\right) - \ln\left((1-\mu_l)\,l_t\right) \\
\varepsilon_{At} &= \ln\left(a_{t+1}\right) - \ln\left(p_t e^{-c_{pa}a_t} + (1-\mu_a)\,a_t\right).
\end{aligned}
$$

(There is a computational caveat here: If a state variable is zero, the
program that computes the log residuals will crash. You will need to
replace the zeros with a positive number less than one, for example 0.5.)
For the demographic noise LPA model, we would compute the *square*

root residuals. In general, the residuals on the transformed scale would be

$$
\begin{aligned}
\varepsilon_{Lt} &= \phi\left(l_{t+1}\right) - \phi\left(ba_t e^{-c_{el}l_t - c_{ea}a_t}\right) \\
\varepsilon_{Pt} &= \phi\left(p_{t+1}\right) - \phi\left((1 - \mu_l)\, l_t\right) \\
\varepsilon_{At} &= \phi\left(a_{t+1}\right) - \phi\left(\left(p_t e^{-c_{pa}a_t} + (1 - \mu_a)\, a_t\right)\right).
\end{aligned}
$$

Note that the residuals ε_{it} themselves are realizations of the random variables $\sigma_i E_{it}$ in the stochastic models.

Because each parameter appears in only one equation and we are assuming no covariances, the likelihood function is the product

$$
\mathcal{L} = \left(\prod \frac{1}{\sigma_L \sqrt{2\pi}} e^{-\frac{\varepsilon_{Lt}^2}{2\sigma_L^2}}\right) \left(\prod \frac{1}{\sigma_P \sqrt{2\pi}} e^{-\frac{\varepsilon_{Pt}^2}{2\sigma_P^2}}\right) \left(\prod \frac{1}{\sigma_A \sqrt{2\pi}} e^{-\frac{\varepsilon_{At}^2}{2\sigma_A^2}}\right),
$$

$$(6.7)$$

assuming the residuals are normally distributed. As in Chapter 2, maximizing the likelihood in this case is equivalent to using the method of CLS to parameterize the model. One nice thing to know about CLS is that it loosens the restrictive assumptions on the residuals; CLS parameter estimates converge to the true values even if the noise is non-normal and autocorrelated, as long as the noise has a stationary distribution (Cushing et al. 2003; Tong 1990).

6.7.2 Conditioned Least Squares (CLS)

Suppose the data time series is of duration $t = 0, 1, 2, \ldots, q$. The residual sums of squares are

$$
\text{RSS}_L = \sum_{t=0}^{q-1} \varepsilon_{Lt}^2
$$

$$
\text{RSS}_P = \sum_{t=0}^{q-1} \varepsilon_{Pt}^2
$$

$$
\text{RSS}_A = \sum_{t=0}^{q-1} \varepsilon_{At}^2.
$$

Thus, for the environmental noise LPA model, we have

$$\text{RSS}_L(b, c_{el}, c_{ea}) = \sum_{t=0}^{q-1} \left[\ln(l_{t+1}) - \ln\left(ba_t e^{-c_{el}l_t - c_{ea}a_t}\right) \right]^2$$

$$\text{RSS}_P(\mu_l) = \sum_{t=0}^{q-1} \left[\ln(p_{t+1}) - \ln\left((1 - \mu_l) l_t\right) \right]^2$$

$$\text{RSS}_A(c_{pa}, \mu_a) = \sum_{t=0}^{q-1} \left[\ln(a_{t+1}) - \ln\left(p_t e^{-c_{pa}a_t} + (1 - \mu_a) a_t\right) \right]^2.$$

Because each parameter appears in only one equation, the three RSS values are minimized if and only if the sum

$$\text{RSS}(b, c_{el}, c_{ea}, \mu_l, c_{pa}, \mu_a) = \text{RSS}_L(b, c_{el}, c_{ea}) + \text{RSS}_P(\mu_l) + \text{RSS}_A(c_{pa}, \mu_a)$$

is minimized as a function of the six parameters. The minimizing parameters are called the *conditioned least squares* (CLS) parameter estimates. The estimated variances are given by

$$\widehat{\sigma}_L^2 = \frac{\widehat{\text{RSS}_L}}{q}; \ \widehat{\sigma}_P^2 = \frac{\widehat{\text{RSS}_P}}{q}; \ \widehat{\sigma}_A^2 = \frac{\widehat{\text{RSS}_A}}{q},$$

where $\widehat{\text{RSS}}$ denotes the fitted value of RSS.

The parameters for the environmental noise LPA model and the environmental noise LPAalt model, estimated from the data in Data Set 6.1 and reported to four significant figures, are listed in Tables 6.1–6.4. In Exercise 5, you will complete Tables 6.1–6.4 by parameterizing the demographic noise models.

Note that the estimated values of the six parameters held in common by the environmental noise LPA and LPAalt models are similar; in fact, to four significant figures, they differ only in the values

TABLE 6.1 LPA Model Parameters Estimated from Data Set 6.1

LPA model	$\widehat{b}$	$\widehat{c}_{el}$	$\widehat{c}_{ea}$	$\widehat{\mu}_l$	$\widehat{c}_{pa}$	$\widehat{\mu}_a$
Environmental noise	10.57	0.009463	0.009649	0.5129	0.01817	0.1065
Demographic noise						

TABLE 6.2 LPA Model Parameters Estimated from Data Set 6.1

LPA model	$\widehat{RSS_L}$	$\widehat{RSS_P}$	$\widehat{RSS_A}$	$\widehat{RSS}$	$\widehat{\sigma}_L^2$	$\widehat{\sigma}_P^2$	$\widehat{\sigma}_A^2$
Environmental noise	20.99	32.56	0.8440	54.39	0.2762	0.4284	0.01111
Demographic noise							

TABLE 6.3 LPAalt Model Parameters Estimated from Data Set 6.1

LPAalt model	$\widehat{b}$	$\widehat{c}_{el}$	$\widehat{c}_{ea}$	$\widehat{\mu}_l$	$\widehat{c}_{pl}$	$\widehat{c}_{pa}$	$\widehat{\mu}_a$
Environmental noise	10.57	0.009463	0.009649	0.5129	0.0007201	0.01782	0.1025
Demographic noise							

TABLE 6.4 LPAalt Model Parameters Estimated from Data Set 6.1

LPAalt model	$\widehat{RSS_L}$	$\widehat{RSS_P}$	$\widehat{RSS_A}$	$\widehat{RSS}$	$\widehat{\sigma}_L^2$	$\widehat{\sigma}_P^2$	$\widehat{\sigma}_A^2$
Environmental noise	20.99	32.56	0.8401	54.39	0.2762	0.4284	0.01105
Demographic noise							

of c_{pa} and μ_a. (This is because the c_{pl} parameter affects only the third equation.) The introduction of the parameter c_{pl} in the LPAalt model slightly decreases the value of c_{pa} and also μ_a. This suggests that there may be some cannibalism of pupae by larvae, but not very much (because c_{pl} is small compared to the other cannibalism coefficients). The question becomes: Is cannibalism of pupae by larvae important enough that we must include it in the model? To answer this question, we next turn to the methods of model selection.

6.8 MODEL SELECTION

We will select the best of the two alternative models (LPA and LPAalt) by using the Akaike information criterion (AIC). *The AIC presupposes that all alternative models under consideration have the same number of state variables and that they are all parameterized on the same data set.* Both of these conditions apply for the LPA and LPAalt models.

The AIC, in its most general formulation, is

$$\text{AIC} = -2\ln\mathcal{L} + 2\kappa, \qquad (6.8)$$

where $\mathcal{L}$ is the maximized value of the likelihood function and κ is the number of estimated parameters in the model, including the variance and covariance parameters. The model with the smallest AIC is the best of the alternatives.

Given assumption (A12) and the fact that each parameter in the deterministic skeleton appears in only one equation, the AIC is equal to (Exercise 4)

$$\text{AIC} = pq(\ln(2\pi) + 1) + q\sum_{i=1}^{p}\ln\left(\widehat{\sigma}_i^2\right) + 2\kappa, \qquad (6.9)$$

where p is the number of state variables ($p = 3$ in our case), q is the number of one-step transitions, the $\widehat{\sigma}_i^2$ are the estimated variances for the state variables, and κ is the number of parameters, which includes the parameters in the skeleton plus the number of variances. For the LPA model, $\kappa = 6 + 3 = 9$, whereas for the LPAalt model, $\kappa = 7 + 3 = 10$.

In Exercise 6, the reader will complete Table 6.5. Here, $\Delta\text{AIC} = \text{AIC} - \text{AIC}_{\min}$. The best model (the one with $\text{AIC}_{\min}$) will have $\Delta\text{AIC} = 0$, and the second-best model will have a value of ΔAIC greater than zero.

In completing Table 6.5 (Exercise 6), the reader will find that the LPA model is the best of the two alternatives. Thus, although there may be some cannibalism of pupae by larvae, that interaction can and should be ignored in the model. This means that we now discard the LPAalt model and focus our attention on validating the LPA model.

TABLE 6.5　Model Selection Based on Parameters Estimated from Data Set 6.1

Model	AIC	ΔAIC
LPA, environmental noise	160.8	
LPAalt, environmental noise		

6.9 MODEL VALIDATION

To validate a model, we first compute the goodness of fit on estimation data, and then, without re-estimating parameters, we compute the goodness of fit on an independent data set. To measure goodness of fit, we define a generalized R^2 for each state variable (Dennis et al. 2001). For example, for the environmental noise LPA model, the R^2 for state variable L is

$$R_L^2 = 1 - \frac{\sum\limits_{t=0}^{q-1}\left[\ln\left(l_{t+1}\right) - \ln\left(ba_t e^{-c_{el}l_t - c_{ea}a_t}\right)\right]^2}{\sum\limits_{t=0}^{q-1}\left[\ln\left(l_{t+1}\right) - \overline{\ln\left(l\right)}\right]^2},$$

where $\overline{\ln\left(l\right)}$ denotes the sample mean of the log-transformed observed L-stage abundances. A similar R^2 formula holds for P and A. In general, the R^2 for state variable L on the ϕ-scale is

$$R_L^2 = 1 - \frac{\sum\limits_{t=0}^{q-1}\left[\phi\left(l_{t+1}\right) - \phi\left(ba_t e^{-c_{el}l_t - c_{ea}a_t}\right)\right]^2}{\sum\limits_{t=0}^{q-1}\left[\phi\left(l_{t+1}\right) - \overline{\phi\left(l\right)}\right]^2},$$

where $\overline{\phi\left(l\right)}$ denotes the sample mean of the ϕ-transformed observed L-stage abundances. A similar R^2 formula holds for P and A.

Data Set 6.2 contains part of a data set collected in 1993–1994 by R. F. Costantino when he was on sabbatical at the lab of Alan Hastings at the University of California at Davis. In this experiment, the adult death rate μ_a was manipulated at the values 0.04, 0.27, 0.5, 0.73, and 0.96. Data Set 6.2 contains the control and the treatments for $\mu_a = 0.5$ and $\mu_a = 0.96$. The entire data set is listed in Appendix B of the book by Cushing et al. (2003).

Table 6.6 gives the R^2 values for the estimation data in Data Set 6.1 and the R^2 values (without refitting) for the independent data in Data Set 6.2 (Exercise 8). *Note that we do not compute validation R^2 for the LPAalt model, because that model has already been discarded.*

A comparison of the R^2 values in Table 6.6 suggests that the LPA model under the assumption of environmental noise is not well validated for the L-stage. This is not surprising, because collection of the estimation and validation data sets occurred over a decade apart

TABLE 6.6   Model Validation

Environmental Noise LPA Model	Estimation R^2	Validation R^2
L	0.80	0.42
P		
A		

Parameters were estimated on Data Set 6.1. Validation was attempted on Data Set 6.2 without re-estimating parameters.

in different laboratories using different culture protocols and different strains of *Tribolium castaneum*. What is surprising, however, is that the validation R^2 values are positive and quite high for ecological data. In Dennis et al. (1997), the entire data set on which Data Set 6.2 is based was randomly divided into estimation data and validation data. The subsequently well-validated model documented a transition between equilibria (control), 2-cycles ($\mu_a = 0.5$), and aperiodic cycles ($\mu_a = 0.96$) for the treatments in Data Set 6.2 (Exercise 9).

In the next section, we revisit a rigorous historical validation of the LPA model.

6.10   THE "HUNT FOR CHAOS" EXPERIMENTS

Let's back up to the beginning of the story. After Lord Robert May proposed the hypothesis that complex fluctuations in populations could be due to deterministic feedback mechanisms (see Chapter 4), researchers set out to search for chaos in population time series data. Most of the data sets were from field systems (Cushing et al. 2003). The multidisciplinary Beetle Team took a different approach. If they located a biologically feasible region in parameter space in which a well-validated model predicted chaotic population dynamics, and placed experimental laboratory populations in that parameter region, would the observed experimental dynamics be chaotic? If one could not demonstrate chaos in laboratory populations, how could one hope to demonstrate it in natural populations? Therefore, they decided that the starting place for the hunt for chaos should be in the laboratory.

Costantino and Desharnais had studied flour beetle dynamics in the lab for many years prior to the Beetle Team collaboration. Based on their thorough understanding of the *Tribolium* system, the Team constructed a mathematical model that incorporated the main mechanisms thought to drive the dynamics. This was the origin of the LPA model. Using historical time series data, the Team parameterized and validated the model (as explained in the previous part of this chapter) and then used the fitted model to make predictions about how the system dynamics would respond to various experimental manipulations of the parameters. Based on the model predictions, they designed and carried out experiments to test these predictions.

Suppose we set the LPA parameters to be the CLS estimates for the demographic LPA model from Dennis et al. (2001): $b = 10.45$, $c_{el} = 0.01731$, $c_{ea} = 0.01310$, $\mu_l = 0.2000$, $c_{pa} = 0.004619$, $\mu_a = 0.007629$. If the experimenter could fix the adult death rate μ_a at 0.96 and manipulate c_{pa} at various values between 0 and 1 in the laboratory, what dynamics would we predict to see in the beetle populations? The easiest way to answer this question is to have the computer draw a bifurcation diagram with c_{pa} as the bifurcation parameter, using $\mu_a = 0.96$ and all other parameter values as listed directly above. The bifurcation diagram for total population size $(L + P + A)$ is shown in Figure 6.3 (Exercise 10).

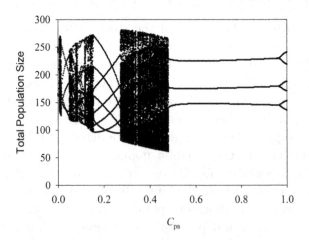

Figure 6.3: Demographic LPA model bifurcation diagram using CLS parameter estimates from Dennis et al. (2001): $b = 10.45$, $c_{el} = 0.01731$, $c_{ea} = 0.01310$, $\mu_l = 0.2000$, $\mu_a = 0.96$. The bifurcation parameter c_{pa} ranges from 0 to 1.

To test the predictions of the bifurcation diagram in Figure 6.3, the Beetle Team designed a way to manipulate μ_a and c_{pa} in the laboratory. They chose seven values of c_{pa} (0.00, 0.05, 0.10, 0.25, 0.35, 0.50, and 1.00) at which to run experimental treatments, replicating each treatment three times. They also had a control with three replicates, in which c_{pa} was not manipulated. The parameter μ_a was manipulated to be 0.96 in all 24 populations. The LPA model, with all other parameters set to those in the caption of Table 6.7, estimated under the assumption of demographic noise, predicts the dynamics shown in Table 6.7. Of particular interest are the transitions between dynamic states, for example, from an equilibrium (control) to chaos ($c_{pa} = 0.35$) to a 3-cycle ($c_{pa} = 0.5$). That is, the LPA model predicts not only attractors, but also *transitions*, as c_{pa} is "tuned" from 0 to 1. The idea was to document transitions and to "bookend" complicated dynamics in the middle range of c_{pa} with simple dynamics at each end of the c_{pa} range.

The plan was for Bob Costantino to count L, P, and A numbers in the 24 populations every two weeks for 80 weeks and then to compare the observed dynamics to the predicted dynamics. At the end of this section, we discuss the general idea of how μ_a and c_{pa} were manipulated in the laboratory; details are given in the papers by Costantino et al. (1997) and Dennis et al. (2001).

TABLE 6.7   LPA Model Predictions. Parameters are CLS estimates for the demographic LPA model from Dennis et al. (2001): $b = 10.45$, $c_{el} = 0.01731$, $c_{ea} = 0.01310$, $\mu_l = 0.2000$, $c_{pa} = 0.004619$, $\mu_a = 0.007629$. Compare with Figure 6.3.

c_{pa}	Prediction with $\mu_a = 0.96$
Control	Equilibrium
0.00	Double invariant loop
0.05	Chaos
0.10	26-cycle
0.25	8-cycle
0.35	Chaos
0.50	3-cycle
1.00	6-cycle

6.10.1 Results of the "Hunt for Chaos" Experiments

The observed dynamics for the "Hunt" experiments are found in Data Set 6.3. For a complete comparison of the data with model-predicted dynamics, see Costantino et al. (1997) and Dennis et al. (2001). Here we show data for the control and three treatments ($c_{pa} = 0.0$, $c_{pa} = 0.35$, and $c_{pa} = 0.5$; Figure 6.4). For the control, we can see that the data are clustered around the model-predicted equilibrium (Figure 6.4a). For $c_{pa} = 0.5$, we see that the data are clustered about the three points of the model-predicted 3-cycle (Figure 6.4d). These are the "bookends." For $c_{pa} = 0.0$, we see that the data have parted into two clusters around the model predicted double invariant loop (Figure 6.4b). For $c_{pa} = 0.35$, the data are strung out along the model-predicted chaotic attractor (Figure 6.4c). The data patterns in each panel are quite different. The strongest evidence that the model is working comes from the fact that it predicted the *transitions* between a simple equilibrium, complicated dynamics, and a simple 3-cycle.

The Beetle Team announced the results of the Hunt for Chaos experiments in *Science* in 1997 (Costantino et al. 1997), followed by a comprehensive paper published in *Ecological Monographs* in 2001 (Dennis et al. 2001). Costantino continued the Hunt for Chaos experiment for 8 years; the data set eventually reached a length of 424 weeks. The Team conducted many more experiments of various types in which they further tested the nonlinear dynamic predictions of the LPA model; you can find these in the annotated references at the end of this chapter. The LPA model has been the most rigorously validated mathematical population model in the history of ecology to date.

6.10.2 Manipulating the Parameters in the Laboratory

Here, we give the general idea of how μ_a and c_{pa} were manipulated in the laboratory; details are given in Dennis et al. (2001).

The adult mortality rate was manipulated to be $\mu_a = 0.96$ and the adult recruitment rate was manipulated so that it would equal $P_t e^{-c_{pa} A_t}$. To manipulate the adult death rate μ_a, at each census Costantino counted dead adults and then removed or added older live adults to the population as necessary. To manipulate c_{pa}, he further removed or added younger adults to make the third equation in model (6.1) exact. For example, at census $t + 1$ for a $c_{pa} = 0.10$ culture, he

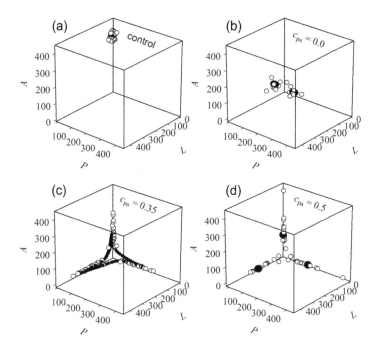

Figure 6.4: Data (open circles) and deterministic model predictions (solid circles) for the LPA model. Parameters are CLS estimates for the demographic noise LPA model from Dennis et al. (2001): $b = 10.45$, $c_{el} = 0.01731$, $c_{ea} = 0.01310$, $\mu_l = 0.2000$, $c_{pa} = 0.004619$, $\mu_a = 0.007629$. (a) Control. Model predicts an equilibrium. (b) $c_{pa} = 0$. Model predicts a double invariant loop. (c) $c_{pa} = 0.35$. Model predicts chaos. (d) $c_{pa} = 0.5$. Model predicts a 3-cycle.

would compute the value of

$$P_t e^{-0.10 A_t} + (1 - 0.96) A_t$$

based on the previous census values of P_t and A_t. This new value was the target value for A_{t+1}. Given the observed number of adults at time $t+1$, Costantino removed or added adults to the culture as needed. Of course, the observed number of adults (not the manipulated number) was the number recorded in the data time series.

When the experimental results were announced in *Science* (Costantino et al. 1997) and the Team began giving talks about the results at research conferences, there was often a question from the audience as to whether "making the third equation exact" was, in fact, driving the dynamics and forcing them to fit the model predictions. This is a very important question. It is right to be skeptical in science, and not only of the work of others. One must also be relentlessly skeptical of one's own work. This was a question that demanded rigorous thinking.

In order to see whether making the third equation of the LPA model exact was forcing the experimental dynamics to fit the model predictions, we consider the following thought experiment. Note that the experimental protocol for manipulating μ_a and c_{pa} depends only on the form of the third equation. Thus, in the lab, you would carry out the exact same experimental protocol in testing *any* larval-pupal-adult model whose third equation was of the form

$$z_{t+1} = y_t e^{-c_{pa} z_t} + (1 - \mu_a) z_t.$$

As an extreme example, we can imagine that one could have been testing the following model (rather than the LPA model):

$$
\begin{aligned}
x_{t+1} &= 0 \\
y_{t+1} &= 0 \\
z_{t+1} &= y_t e^{-c_{pa} z_t} + (1 - \mu_a) z_t.
\end{aligned}
$$

This ridiculous model predicts zero larvae and pupae for all $t > 0$, along with a geometrically decreasing number of adults. Nevertheless, *the experimental results would have been the same as in Figure 6.4!* Clearly, "making the third equation exact" does not force the experimental dynamics to follow the predictions of the model.

Can you think of a situation in which the experimental approach of "making one equation exact" would not be legitimate? See Exercise 11.

6.11 EXERCISES

1. In this problem, you will carry out a stability analysis for the extinction state of the LPA model (6.1).

   a. Prove that the extinction state $(0, 0, 0)$ is an equilibrium for the LPA model.

b. Find the linearization of the LPA model at the equilibrium $(0, 0, 0)$.

c. Find the characteristic equation of the linearization. Hint: If you do not know how to find the determinant of a 3×3 matrix, look up the procedure in an introductory linear algebra text or calculus text.

d. Can you find conditions on the parameters that guarantee the stability/instability of the extinction state $(0, 0, 0)$?

e. If you were able to complete part (1d), define the *inherent net reproductive number* n by $n = b(1 - \mu_l)/\mu_a$. This is the expected number of adults arising from one adult in the absence of cannibalism. Rewrite your conditions from part (1d) in terms of n. Give a biological interpretation of these conditions.

2. Consider a two-dimensional beetle model. Here, we will combine the larvae and pupae into a juvenile stage J_t. The time step is approximately 4 weeks, the sum of the durations of these stages. Assume that (a) the egg stage can be ignored; (b) in the absence of cannibalism, the per capita juvenile recruitment rate is a constant $b > 0$ juveniles per adult per unit time; (c) in the presence of cannibalism, the juvenile recruitment rate is decreased by a Ricker factor of the form $e^{-c_{ej}J}$ due to cannibalism of eggs by larvae; (d) in the presence of cannibalism, the juvenile recruitment rate is further decreased by a Ricker factor of the form $e^{-c_{ea}A}$ due to the cannibalism of eggs by adults; (e) in the absence of cannibalism, there are no juvenile deaths; (f) in the presence of cannibalism, the juvenile survivorship is decreased by a Ricker factor of the form $e^{-c_{ja}A}$ due to the cannibalism of juveniles by adults; (g) a fraction μ_a of adults die per unit time, where $0 < \mu_a < 1$; (h) there is no measurement error; (i) demographic stochasticity is the main source of noise in the system; and (j) stochastic perturbations are uncorrelated across life-cycle stages and uncorrelated in time. Let

$$J_t = \text{Number of larvae and pupae at time } t$$
$$A_t = \text{Number of adults at time } t.$$

  a. Draw the Leslie Diagram for the JA model.

  b. Write the equations for the deterministic JA model. Always state the time step when you pose a discrete-time model.

  c. Write the equations for the stochastic JA model assuming demographic noise.

3. Write equations for the "environmental noise LPAalt model" and the "demographic noise LPAalt model."

4. Derive equation (6.9) from equations (6.7) and (6.8).

5. In this problem, you will parameterize the two alternative models presented in this chapter, the LPA and LPAalt models, using the data in Data Set 6.1 as the estimation data.

  a. Reproduce all the values in Tables 6.1–6.4 for the environmental noise models.

  b. Complete Tables 6.1–6.4 for the demographic noise models.

6. In this problem, you will use the AIC to select the best of the two alternatives, the LPA and LPAalt models, under the assumption of environmental noise.

  a. Using equation (6.9), complete Table 6.5.

  b. Select the best model and explain your selection.

7. In this problem, you will explore the dynamics of the parameterized LPA model and make testable predictions. Use the parameters estimated from the environmental noise LPA model (Tables 6.1–6.2).

  a. Write a program to simulate the deterministic LPA model. Produce a time series of length 40, given the initial condition $(250, 5, 100)$. Plot the time series for L, P, and A on the same graph. Attach (i) your program, (ii) the time series (as a matrix with 40 rows and 3 columns), and (iii) the graph.

b. Write a program to simulate the stochastic LPA model with environmental noise (6.5). Produce a typical stochastic time series of length 40, given the initial condition $(250, 5, 100)$. Plot the time series for L, P, and A on the same graph. Attach (i) your program, (ii) the time series (as a matrix with 40 rows and 3 columns), and (iii) the graph.

c. Use the computer to graph the LPA model bifurcation diagram for larval numbers, using μ_a as a bifurcation parameter with range $0 \le \mu_a \le 1$.

d. The bifurcation diagram in (7c) is a prediction about the behavior of the flour beetle dynamics as a function of the adult death rate μ_a. What dynamic does the model predict for $\mu_a = 0.1065$? For $\mu_a = 0.50$? For $\mu_a = 0.96$? Try to answer these questions simply by looking at the bifurcation diagram.

e. Imagine planning a laboratory experiment to test the predictions in part (7d). Your design would involve a control (in which μ_a is not manipulated) and perhaps two treatments in which μ_a is manipulated at the fixed values $\mu_a = 0.5$ and $\mu_a = 0.96$, with replicates of the control and each treatment. In order to further explore the predicted outcomes:

    i. Prepare time series graphs that show the deterministic model predictions for each of your experimental treatments and the control for 40 time steps using the initial condition $(250, 5, 100)$. Attach graphs of time series for L and A.

    ii. Prepare time series graphs that show typical stochastic model simulations for each of your experimental treatments and the control for 40 time steps using the initial condition $(250, 5, 100)$. Attach graphs of time series for L and A.

8. In this problem, you will check to see if the LPA model fits Data Set 6.2 without re-estimating the parameters. Use the parameters estimated from the environmental noise LPA model (Tables 6.1–6.2). Note that the experimental manipulations began at week 12 (Data Set 6.2).

a. Fill in the R^2 values for the estimation data (Data Set 6.1) in Table 6.6.

b. *Without re-estimating the model parameters*, compute the goodness of fit for the data in Data Set 6.2. That is, fill in the R^2 values for the validation data in Table 6.6. Hint: This exercise will require some careful thought and programming. There are several issues to keep in mind. First, look at Data Set 6.2 and note that there is a "split" in each treatment. Weeks 0–12 did not undergo manipulation. At week 12, the experiment was "restarted" using the week 12 observed values as initial conditions and applying the manipulations. This is why week 12 is listed twice in Data Set 6.2 for each treatment. Second, you will use the estimated value of $\mu_a = 0.1065$ not only for the unmanipulated control, but also for the initial 12 weeks of each of the other two treatments. For the remaining weeks of the manipulated treatments, you will set $\mu_a = 0.5$ or $\mu_a = 0.96$, as appropriate. Third, when creating residuals, it is important to correctly match up predictions with observations with these "splits" in mind. Fourth, note that Data Set 6.2 contains some values of zero, which will cause problems with log residuals. When computing the log of the state variables, replace values of 0 with 0.5.

c. A comparison of the R^2 values in Table 6.6 suggests that the LPA model under the assumption of environmental noise is not well validated for the L-stage. This is not surprising, because the collection of the estimation and validation data sets occurred over a decade apart in different laboratories using different culture protocols and different strains of *Tribolium castaneum*. In Dennis et al. (1997) the entire data set on which Data Set 6.2 is based was randomly divided into estimation data and validation data. The subsequently well-validated model documented a transition between equilibria (control), 2-cycles ($\mu_a = 0.5$), and aperiodic cycles ($\mu_a = 0.96$) for the treatments in Data Set 6.2 (Exercise 9). You can see these transitions if you look at the time series in Data Set 6.2. In this current exercise, however, continue to use the parameters estimated from the environmental noise LPA model in Tables 6.1–6.2. Systematically perturb the parameters and rerun the bifurcation diagram in (7c) to explore how it changes. Can you find parameters "nearby" in parameter space that produce a transition from an equilibrium to 2-cycles to aperiodic cycles as μ_a is "tuned" from 0 to 1?

9. Read Dennis et al. (1997). We will use the parameters estimated for the SS strain in this problem: $b = 7.483$, $c_{el} = 0.01200$, $c_{ea} = 0.009170$, $\mu_l = 0.2670$, $c_{pa} = 0.004139$, $\mu_a = 0.003620$.

   a. Use the computer to graph the LPA model bifurcation diagram for larval numbers, using μ_a as a bifurcation parameter with range $0 \leq \mu_a \leq 1$.

   b. Repeat the validation procedure in problem (8b) using these parameters.

10. Read Costantino et al. (1997) and Dennis et al. (2001). Reproduce the LPA bifurcation diagram in Figure 6.3 using the parameters shown in the figure caption.

11. Suppose you are testing the predictions of a one-dimensional Ricker model

$$x_{t+1} = bx_t e^{-cx_t}$$

for fish in the laboratory, with time step of one year, and you want to manipulate the parameters as follows. Suppose you want to manipulate the parameter c to be $c = 0.01$ and you want to set up experimental cultures with $b = 0.5$, $b = 5$, and $b = 10$.

   a. Draw the bifurcation diagram, using b as a bifurcation parameter for $0 \leq b \leq 12$.

   b. What are the model-predicted transitions as b is tuned from 0 to 12? What are the predicted dynamics for each of $b = 0.5$, $b = 5$, and $b = 10$?

   c. Would it be legitimate to test the Ricker predictions in the laboratory by manipulating b and c through adding or removing fish at each time step (after the observed census was recorded) in order to "make the Ricker equation exact"?

Stated negatively, does that experimental protocol force the observed dynamics to correspond to the predicted Ricker dynamics? Explain your reasoning.

BIBLIOGRAPHY

Costantino, R. F., Cushing, J. M., Dennis, B., and Desharnais, R. A. 1995. Experimentally induced transitions in the dynamic behavior of insect populations. *Nature* 375:227–230. DOI: 10.1038/375227a0. ["Announcement" paper with first appearance of the LPA model in the literature. Documents model-predicted bifurcations from stable fixed points to periodic cycles to aperiodic oscillations through experimental manipulation of the adult mortality rate in the laboratory. Both theoretical and empirical: models connected to data. The follow-up paper by Dennis et al. (1997) gives the details.]

Costantino, R. F., Cushing, J. M., Dennis, B., Desharnais, R. A., and Henson, S. M. 1998. Resonant population cycles in temporally fluctuating habitats. *Bulletin of Mathematical Biology* 60:247–273. DOI: 10.1006/bulm.1997.0017. [LPA model predicts, under certain circumstances, a larger average total population abundance when the habitat volume periodically fluctuates than when the habitat volume is held constant at the average volume. Tested this prediction empirically in laboratory populations. Both theoretical and empirical: models connected to data.]

Costantino, R. F. and Desharnais, R. A. 1991. *Population Dynamics and the Tribolium Model: Genetics and Demography.* Springer-Verlag, New York. [Provides a comprehensive history of the mathematics of *Tribolium*.]

Costantino, R. F., Desharnais, R. A., Cushing, J. M., and Dennis, B. 1997. Chaotic dynamics in an insect population. *Science* 275:389–391. DOI: 10.1126/science.275.5298.389. ["Announcement" paper heralding the documentation of chaos in a laboratory insect population. Both theoretical and empirical: models connected to data. The follow-up paper by Dennis et al. (2001) gives the details.]

Costantino, R. F., Desharnais, R. A., Cushing, J. M., Dennis, B., Henson, S. M., and King, A. A. 2005. The flour beetle *Tribolium* as an effective tool of discovery. *Advances in Ecological Research* 37:101–141. DOI: 10.1016/S0065-2504(04)37004-2. [Summary of the methodology of the Beetle Team, emphasizing that both deterministic and stochastic forces are important in ecological systems. Discusses how a number of nonlinear phenomena have been documented in a laboratory system, including chaotic dynamics, population outbreaks, saddle nodes, phase switching, and lattice effects. Also discusses the anatomy of chaos and mechanistic models of stochasticity. Both theoretical and empirical: models connected to data.]

Cushing, J. M., Costantino, R. F., Dennis, B., Desharnais, R. A., and Henson, S. M. 1998. Nonlinear population dynamics: Models, experiments, and data. *Journal of Theoretical Biology* 194:1–9. DOI: 10.1006/jtbi.1998.0736. ["Perspectives" piece on the research paradigm of the Beetle Team. Both theoretical and empirical: models connected to data.]

Cushing, J. M., Costantino, R. F., Dennis, B., Desharnais, R. A., and Henson, S. M. 2003. *Chaos in Ecology: Experimental Nonlinear Dynamics.* Academic Press, San Diego, CA. [Summarizes the large body of work involved in the documentation of chaos in a laboratory population. Includes several data sets in the appendices.]

Cushing, J. M., Dennis, B., Desharnais, R. A., and Costantino, R. F. 1996. An interdisciplinary approach to understanding nonlinear ecological dynamics. *Ecological Modelling* 92:111–119. DOI: 10.1016/0304-3800(95)00170-0. [Describes a methodology for testing nonlinear population theory using data, models, experiments, and statistics. Both theoretical and empirical: models connected to data.]

Cushing, J. M., Dennis, B., Desharnais, R. A., and Costantino, R. F. 1998. Moving toward an unstable equilibrium: saddle nodes in population systems. *Journal of Animal Ecology* 67:298–306. DOI: 10.1046/j.1365-2656.1998.00194.x. [Empirical demonstration of a model-predicted "saddle fly-by" in *Tribolium* data, in which the system approached an unstable equilibrium along a two-dimensional stable manifold and departed along a one-dimensional unstable manifold. Both theoretical and empirical: models connected to data.]

Cushing, J. M., Henson, S. M., Desharnais, R. A., Dennis, B., Costantino, R. F., and King, A. 2001. A chaotic attractor in ecology: Theory and experimental data. *Chaos, Solitons, and Fractals* 12:219–234. DOI: 10.1016/S0960-0779(00)00109-0. [The LPA chaotic attractor and the stochastic LPA model's stationary distribution are compared in detail to the experimental data in phase space and are showed to correspond to an amazing degree. Model-predicted temporal patterns on the chaotic attractor are demonstrated in the data, as well. Both theoretical and empirical: models connected to data.]

Dennis, B., Desharnais, R. A., Cushing, J. M. and Costantino, R. F. 1995. Nonlinear demographic dynamics: mathematical models, statistical methods, and biological experiments. *Ecological Monographs* 65:261–281. DOI: 10.2307/2937060. [Introduces the LPA model and analyzes the Desharnais data, the controls of which appear in Data Set 6.1.]

Dennis, B., Desharnais, R. A., Cushing, J. M., and Costantino, R. F. 1997. Transitions in population dynamics: equilibria to periodic cycles to aperiodic cycles. *Journal of Animal Ecology* 66:704–729. DOI: 10.2307/5923. [Detailed follow-up to the "announcement" paper in *Nature* (Costantino et al. 1997). From the abstract: "The rigorous statistical verification of the predicted shifts in dynamical behaviour provides convincing evidence for the relevance of nonlinear mathematics in population biology." Both theoretical and empirical: models connected to data.]

Dennis, B., Desharnais, R. A., Cushing, J. M., Henson, S. M., and Costantino, R. F. 2001. Estimating chaos and complex dynamics in an insect population. *Ecological Monographs* 71:277–303. DOI: 10.1890/0012-9615(2001)071[0277:ECACDI]2.0.CO;2. [Detailed follow-up to the "announcement" paper in *Science* (Costantino et al. 1997).]

Desharnais, R. A. and Costantino, R. F. 1980. Genetic analysis of a population of *Tribolium*. VII. Stability: Response to genetic and demographic perturbations. *Canadian Journal of Genetics and Cytology* 22:577–589. DOI: 10.1139/g80-063. [First appearance in the literature of the data in Data Set 6.1. Both theoretical and empirical: models connected to data.]

Desharnais, R. A., Costantino, R. F., Cushing, J. M., and Dennis, B. 1997. Estimating chaos in an insect population. *Science* 276:1881–1882. DOI: 10.1126/science.276.5320.1881. [Part of a discussion in *Science* about positive stochastic Lyapunov exponents and whether they indicate chaos in noisy data, motivated by Costantino et al. (1997).]

Desharnais, R. A., Costantino, R. F., Cushing, J. M., Henson, S. M., Dennis, B., and King, A.A. 2006. Experimental support for the scaling rule of demographic stochasticity. *Ecology Letters* 9:537–547. DOI: 10.1111/j.1461-0248.2006.00903.x. [Experimental test of the ecological tenet that the magnitude of demographic-stochastic fluctuations in populations scales inversely with the square root of population size. We experimentally verified the scaling rule for populations displaying noisy chaotic dynamics. Deterministic dynamics were clarified in larger populations. Both theoretical and empirical: models connected to data.]

Desharnais, R. A., Dennis, B., Cushing, J. M., Henson, S. M., and Costantino, R. F. 2001. Chaos and population control of insect outbreaks. *Ecology*

Letters 4:229–235. DOI: 10.1046/j.1461-0248.2001.00223.x. [LPA-model-guided laboratory protocol for applying small perturbations to *Tribolium* populations in order to control chaotic dynamics. Both theoretical and empirical: models connected to data.]

Dong, Q. and Polis, G. A. 1992. The dynamics of cannibalistic populations: A foraging perspective. In M. A. Elgar & B. J. Crespi (Eds.) *Cannibalism: Ecology and Evolution among Diverse Taxa* (pp. 13–37). Oxford University Press, Oxford. [Classic reference on cannibalism.]

Elgar, M. A. and Crespi, B. J. 1992. Ecology and evolution of cannibalism. In M. A. Elgar & B. J. Crespi (Eds.) *Cannibalism: Ecology and Evolution among Diverse Taxa* (pp. 1–12). Oxford University Press, Oxford. [Classic reference on cannibalism.]

Fox, L. R. 1975. Cannibalism in natural populations. *Annual Review of Ecology and Systematics* 6:87–106. DOI: 10.1146/annurev.es.06.110175.000511. [Classic reference on cannibalism.]

Henson, S. M. 2000. Multiple attractors and resonance in periodically-forced population models. *Physica D: Nonlinear Phenomena* 140:33–49. DOI: 10.1016/S0167-2789(99)00231-6. [Mathematical treatment of Henson et al. (1999). Both theoretical and empirical: models connected to data.]

Henson, S. M., Costantino, R. F., Cushing, J. M., Dennis, B., and Desharnais, R. A. 1999. Multiple attractors and population dynamics in periodic habitats. *Bulletin of Mathematical Biology* 61:1121–1149. DOI: 10.1006/bulm.1999.0136. [Populations that oscillate can develop multiple attractors in the advent of periodic forcing. LPA-model predictions include multiple attractors, resonance and attenuation phenomena, and saddle fly-bys. Both theoretical and empirical: models connected to data.]

Henson, S. M., Costantino, R. F., Cushing, J. M., Desharnais, R. A., Dennis, B., and King, A. A. 2001. Lattice effects observed in chaotic dynamics of experimental populations. *Science* 294:602–605. DOI: 10.1126/science.1063358. [Chaotic attractors live in continuous phase space, that is, they take on a continuum of values. Populations live on a lattice in phase space; a bounded population must take on discrete values on a finite lattice. The stochastic LPA model, when confined to a lattice in phase space, shows fragments of signature cycles. These model-predicted signatures are verified in the "Hunt for Chaos" empirical data set. Both theoretical and empirical: models connected to data.]

Henson, S. M., Costantino, R. F., Desharnais, R. A., Cushing, J. M., and Dennis, B. 2002. Basins of attraction: population dynamics with two stable 4-cycles. *Oikos* 98:17–24. DOI: 10.1034/j.1600-0706.2002.980102.x. [Focuses on LPA-model-predicted multiple attractors and basins of attraction, and verifies these phenomena in empirical population data. Both theoretical and empirical: models connected to data.]

Henson, S. M. and Cushing, J. M. 1997. The effect of periodic habitat fluctuations on a nonlinear insect population model. *Journal of Mathematical Biology* 36:201–226. DOI: 10.1007/s002850050098. [LPA model illustrates how a periodically fluctuating environment can increase population numbers.]

Henson, S. M., Cushing, J. M., Costantino, R. F., Dennis, B., and Desharnais, R. A. 1998. Phase switching in population cycles. *Proceedings of the Royal Society, London B* 265:2229–2234. DOI: 10.1098/rspb.1998.0564. [Oscillation phase switches in population cycles are explained as stochastic jumps between basins of multiple attractors. Both theoretical and empirical: models connected to data.]

King, A. A., Costantino, R. F., Cushing, J. M., Henson, S. M. Desharnais, R. A., and Dennis, B. 2003. Anatomy of a chaotic attractor: Subtle model predicted patterns revealed in population data *Proceedings of the National Academy of Sciences* 408–413. DIO: 10.1073/pnas.2237266100. [From the abstract: "By using data drawn from chaotic insect populations, we show quantitatively that chaos manifests itself as a tapestry of identifiable and predictable patterns woven together by stochasticity." Both theoretical and empirical: models connected to data.]

King, A. A., Desharnais, R. A., Henson, S. M., Costantino, R. F., Cushing, J. M., and Dennis, B. 2002. Random perturbations and lattice effects in chaotic population dynamics: Reply to Domokos and Scheuring. *Science* 297:2163a. [Part of a discussion in *Science* about lattice effects in noisy population data, motivated by Henson et al. (2001).]

Polis, G. A. 1981. The evolution and dynamics of intraspecific predation. *Annual Review of Ecology and Systematics* 12:225–251. DOI: 10.1146/annurev.es.12.110181.001301. [Classic paper on cannibalism.]

Sokoloff, A. 1972. *The Biology of Tribolium Vol 1: With Special Emphasis on Genetic Aspects.* Oxford University Press, Oxford. [Comprehensive reference for the life history of flour beetles.]

Sokoloff, A. 1975. *The Biology of Tribolium Vol 2: With Special Emphasis on Genetic Aspects.* Oxford University Press, Oxford. [Comprehensive reference for the life history of flour beetles.]

Sokoloff, A. 1978. *The Biology of Tribolium Vol 3: With Special Emphasis on Genetic Aspects.* Oxford University Press, Oxford. [Comprehensive reference for the life history of flour beetles.]

Tong, H. 1990. *Non-Linear Time Series: A Dynamical System Approach.* Oxford University Press, Oxford. [Valuable introduction to the analysis of nonlinear time series.]

III

Continuous-Time Models

CHAPTER 7

Introduction to Differential Equations

7.1 WHAT YOU SHOULD KNOW ABOUT THIS CHAPTER

Suppose $x(t)$ is the state of a dynamical system at time t. If the system involves mainly continuous-time processes, our model equations would likely involve the continuous-time rate of change of x, which is the derivative dx/dt. Equations involving derivatives are called *differential equations*. Thus, continuous-time processes are often modeled with differential equations.

Undergraduate mathematics majors typically take a course in differential equations after the calculus sequence. In that class, they learn to solve many types of differential equations. Many undergraduate biology majors, however, cannot fit the course into their schedules and proceed to graduate or professional school without having learned how to solve differential equations. Although it would be best if all biology majors took a differential equations course, it is important for biologists, and particularly ecologists, to realize that they can understand how to model with differential equations and they can do much of the most important analysis without ever solving the equations. We will focus on such techniques in Chapter 8.

The current chapter is an introduction to compartmental differential equations and is composed of three examples that illustrate important ideas: (i) an example of how to set up a continuous-time compartmental model using dimensional analysis; (ii) an example of how to study solutions of compartmental models without solving the equations; and (iii) an example of using a powerful solution method to solve a famous differential equation in ecology. We will briefly introduce two techniques (partial fractions and separation of variables) that you would see in second-semester calculus and differential equations classes, but we will explain as we go for those who have not had those courses.

7.2   COMPARTMENTAL MODELS

For compartmental models, the main modeling technique is to look at the net flux, which we can write as

$$\frac{dx}{dt} = \text{Inflow rate} - \text{Outflow rate.}$$

Let's illustrate this idea with three examples.

7.2.1   A Tank Problem

The first example is a generic "tank" model that illustrates inflow and outflow rates as well as the usefulness of thinking in terms of dimensional analysis. Suppose two 100 L tanks are initially full of pure water. Salt solution of concentration 20 g/L flows into tank A at a rate of 3 L/min. The well-mixed solution flows into tank B at a rate of 3 L/min, and tank B is allowed to overflow. Our goal is to model the amount of salt in each tank (Figure 7.1).

Because we want to model the *amount* (in grams) of salt in each tank, we let $x(t)$ and $y(t)$ be the *number of grams* of salt in tank A and B, respectively, at time t. Time t is in minutes, so the derivatives dx/dt and dy/dt will have units of g/min. Thus, each term in the compartmental model—that is, each inflow rate and outflow rate—must also have units of g/min. In order to obtain these fluxes in units of g/min, we must multiply each flow rate of liquid (L/min)

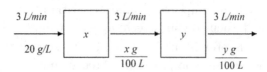

Figure 7.1: Compartments and flows for a system of tanks.

by its salt concentration (g/L). An easy way to see this is that (L/min)(g/L)=g/min because the liter units in each fraction cancel.

The initial amounts of salt in each tank are zero, so the initial conditions are $x(0) = 0$ and $y(0) = 0$. Furthermore, the tanks always remain full of 100 L of solution. However, the concentrations in both tanks change over time. We know the concentration of the inflow into tank A is a constant 20 g/L. The concentration of the outflow from tank A (= the inflow to tank B) will be x *grams* of salt in tank A per 100 L of water in tank A, that is, $x/100$ with units of g/L. A similar consideration yields the outflow from tank B as $y/100$ with units g/L (Figure 7.1).

The compartmental differential equations are

$$\frac{dx}{dt} = \left(3\frac{L}{min}\right)\left(20\frac{g}{L}\right) - \left(3\frac{L}{min}\right)\left(\frac{x}{100}\frac{g}{L}\right)$$

$$\frac{dy}{dt} = \left(3\frac{L}{min}\right)\left(\frac{x}{100}\frac{g}{L}\right) - \left(3\frac{L}{min}\right)\left(\frac{y}{100}\frac{g}{L}\right).$$

The initial value problem is therefore

$$\frac{dx}{dt} = 60 - \frac{3x}{100}$$

$$\frac{dy}{dt} = \frac{3x}{100} - \frac{3y}{100}$$

$$x(0) = 0$$

$$y(0) = 0.$$

7.2.2 The SIR Model

The second example is a famous one from epidemiology, published by Kermack and McKendrick (Kermack and McKendrick 1927). The model is called an *SIR model* (for susceptibles/infectives/recovereds), and it models a disease with rapid spread and permanent acquired immunity. Many models in epidemiology are based on the SIR model.

Let S be the class of *susceptibles* (those who can catch, but not spread the disease). Let I be the class of *infectives* (those who can infect susceptibles). Let R be the class of *recovereds* (those who are recovered with permanent immunity). Let's assume that

(A) $S + I + R$ is constant (so that the time derivatives $S' + I' + R' = 0$).

(B) The per capita rate at which S are infected is proportional to I.

(C) No new S enter the population.

(D) The per capita rate at which I recover is constant.

Given assumption (B), the per capita flow rate from compartment S to compartment I is βI for some $\beta > 0$, and so the total flow rate from S to I is $(\beta I)(S)$. The per capita flow rate from I to R is constant $\gamma > 0$, and so the total flow rate from I to R is γI (Figure 7.2). The SIR model is

$$
\begin{aligned}
\frac{dS}{dt} &= -\beta SI \\
\frac{dI}{dt} &= \beta SI - \gamma I \\
\frac{dR}{dt} &= \gamma I \\
S(0) &= S_0 > 0 \\
I(0) &= I_0 > 0 \\
R(0) &= R_0 > 0,
\end{aligned}
\tag{7.1}
$$

where $\beta, \gamma > 0$ are parameters.

Several interesting implications arise from these model equations. For example, the number of susceptibles $S(t)$ is decreasing as long as there are both susceptibles and infectives in the population, since its derivative is negative. It can be shown, in fact, that $S(t)$ tends to zero in the long run. This means that the derivative of $I(t)$ must eventually go negative, as well, and so $I(t)$ must ultimately decrease. In fact, it can be shown that $I(t)$ also tends to zero in the long run. This means the disease ultimately dies out. However, $I(t)$ can initially increase, before it begins decreasing. If this happens, we say there is an *epidemic*, or

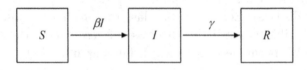

Figure 7.2: Compartments and per capita flow rates for the SIR model.

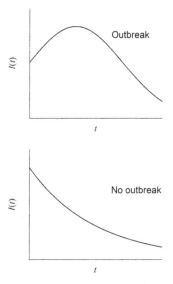

Figure 7.3: If dI/dt is initially positive, we say there is an outbreak.

outbreak. That is, if at time $t = 0$, we have $dI/dt > 0$, then there is an outbreak, whereas if $dI/dt \leq 0$ at $t = 0$, there is no outbreak (Figure 7.3).

Note that $I(t)$ is non-increasing as long as

$$\frac{dI}{dt} = \beta SI - \gamma I \leq 0,$$

that is, as long as

$$S \leq \frac{\gamma}{\beta}.$$

Thus, there will be no outbreak if $S_0 \leq \gamma/\beta$, whereas there will be an outbreak if $S_0 > \gamma/\beta$.

What are the equilibria of the SIR model? The equilibria occur where the derivatives are zero:

$$\frac{dS}{dt} = -\beta SI = 0$$

$$\frac{dI}{dt} = \beta SI - \gamma I = I(\beta S - \gamma) = 0$$

$$\frac{dR}{dt} = \gamma I = 0.$$

From the third equation, we see that the equilibrium value of I must be zero, while S and R can be any non-negative value. So the equilibria all have the form $(S_e, 0, R_e)$, which includes the trivial equilibrium $(0, 0, 0)$. This means there is no *endemic* disease state (i.e., no level of disease that can be maintained at a steady state in the host).

One of the interesting aspects of the SIR example is the amount of information about the system we can obtain without actually solving the differential equations. It is important for biologists to realize that, in general, it is not necessary to actually solve a differential equation in order to study its equilibria and stability. In Chapter 8, we will learn more techniques for analyzing differential equations without solving them. But to finish the current chapter, we illustrate a powerful solution method for solving a famous differential equation from ecology. Hopefully, students who have not taken a course in differential equations will become intrigued and be motivated to do so.

7.2.3   The Continuous-Time Logistic Model

The third and final example in this chapter concerns a famous model from population ecology called the *logistic model* (Verhulst 1838; Verhulst 1845). This model can be posed as a compartmental model

$$\frac{dx}{dt} = bx - dx^2,$$

where $b, d > 0$ are parameters, $x(t)$ is the population size or density at time t, bx is the total population birth rate, and dx^2 is the total population death rate. We will examine the assumptions underlying this model in Chapter 8. For now, we simply note that the logistic model is traditionally written as

$$\frac{dx}{dt} = rx\left(1 - \frac{x}{K}\right), \tag{7.2}$$

where $r = b$ and $K = b/d$ (Exercise 4) (Pearl and Reed 1920). Our goal here is to illustrate how to solve this type of differential equation.

The first step in solving equation (7.2) is to note that there are two equilibrium solutions, namely $x_e = 0$ and $x_e = K$.

The next step is to note that equation (7.2) is multiplicatively *separable*, which means it can be written in the form

$$\frac{dx}{dt} = f(x) g(t).$$ \hfill (7.3)

(In the logistic model, we can take $f(x) = x(1 - x/K)$ and $g(t) = r$.) It turns out that one can legitimately treat dx/dt as a fraction of differentials to multiplicatively "separate the variables" as

$$\frac{dx}{f(x)} = g(t) dt.$$

One can then integrate each side with respect to its variable. It is important to note, however, that dividing by $f(x)$ assumes $f(x) \neq 0$, which "drops" all equilibrium solutions of equation (7.3). In our current example, we can write

$$\frac{dx}{x(1 - x/K)} = rdt$$

as long as $x \neq 0$ and $x \neq K$. So we see that we have "lost" the equilibrium solutions $x_e = 0$ and $x_e = K$.

Then

$$\int \frac{dx}{x(1 - x/K)} = \int rdt + C_1,$$

for $C_1 \in R$, and so

$$\int \frac{dx}{x(1 - x/K)} = rt + C_1.$$

The integrand on the left-hand side is a rational function with linear factors in the denominator. To integrate such a function, we use a technique from second-semester calculus called "partial fraction decomposition." The idea is to decompose the integrand as a sum with unknown coefficients over the linear factors and then determine the coefficients. We therefore decompose the integrand into two fractions and then add them back together:

$$\frac{1}{x(1 - x/K)} = \frac{A}{x} + \frac{B}{(1 - x/K)} = \frac{A(1 - x/K) + Bx}{x(1 - x/K)}.$$

Then, ignoring for a moment the intermediate step, we set the numerators equal to each other:

$$\begin{aligned} 1 &= A\left(1 - x/K\right) + Bx \\ &= A + \left(B - A/K\right)x. \end{aligned}$$

Setting "like coefficients" on x to be equal, we obtain

$$1 = A \text{ and } B - A/K = 0,$$

which leads to $A = 1$ and $B = 1/K$. Thus, using the partial fraction decomposition, we can write

$$\int \left[\frac{1}{x} + \frac{1/K}{(1 - x/K)}\right] dx = rt + C_1.$$

Integration yields

$$\ln|x| - \ln|1 - x/K| = rt + C_1.$$

By a property of logarithms, we have

$$\ln\left|\frac{x}{1 - x/K}\right| = rt + C_1.$$

To solve for x, we exponentiate both sides and then drop the absolute values by introducing $\pm$ on the right:

$$\begin{aligned} \left|\frac{x}{1 - x/K}\right| &= e^{rt+C_1} = e^{C_1}e^{rt} \\ \frac{x}{1 - x/K} &= \pm e^{C_1}e^{rt} = Ce^{rt}. \end{aligned}$$

Note that $C = \pm e^{C_1}$ can be any number except 0. Multiplication of both sides by the denominator yields

$$x = Ce^{rt}\left(1 - x/K\right),$$

and solving for x yields

$$x = \frac{Ce^{rt}}{1 + Ce^{rt}/K} \text{ for } C \neq 0. \tag{7.4}$$

Equation (7.4) is (well, almost) the *general solution*, meaning the set of all possible solutions as indexed by the constant C. But it is NOT the set of all possible solutions, because our algebraic procedure dropped the two solutions $x_e = 0$ and $x_e = K$. We must include these in the general solution, so we write

$$x(t) = \begin{cases} \frac{Ce^{rt}}{1+Ce^{rt}/K} & \text{for } C \neq 0 \\ 0 & \\ K & \end{cases}.$$

Note that if C *were* allowed to be 0, we would pick up the equilibrium solution $x_e = 0$ as part of our solution formula. So why bother with keeping track of $C \neq 0$ and the lost equilibrium solutions throughout the algebraic procedure if, in fact at the end, we allow $C = 0$? The answer is that setting $C = 0$ does not necessarily absorb all the equilibrium solutions into the formula. In this example, it is impossible to obtain the solution $x_e = K$ from the solution formula no matter what you choose for C (Exercise 5). Therefore, the true general solution, denoted x_g, can be written as

$$x_g(t) = \begin{cases} \frac{Ce^{rt}}{1+Ce^{rt}/K} & \text{for } C \in R \\ K & \end{cases}. \tag{7.5}$$

A solution such as $x_e = K$ that cannot be "fit" into the solution formula is called a *singular solution*.

Let the initial condition be $x(0) = x_0$, where $x_0 \neq K$. Applying this initial condition to equation (7.5) obtains a relationship between x_0 and C:

$$x_0 = \frac{KC}{K+C}. \tag{7.6}$$

Solving for C yields (Exercise 6)

$$C = \frac{Kx_0}{K-x_0}. \tag{7.7}$$

Equations (7.5) and (7.7) together yield the *particular solution*, meaning the solution that satisfies the initial condition. The particular solution is (Exercise 7)

$$x(t) = \frac{Kx_0}{x_0 + (K-x_0)e^{-rt}}. \tag{7.8}$$

In Chapter 8, we will study the behavior of these solutions of the logistic equation as they depend on the initial condition x_0 in the context of ecology.

7.3   EXERCISES

1. Consider two 100 L tanks. Tank A is initially full of pure water, and tank B initially contains 50 L of pure water. Seawater of concentration 35 g/L flows into tank A at a rate of 2 L/min. The well-mixed solution overflows into tank B, and when tank B is full, it is allowed to overflow as well. Write the differential equation initial value problem for the amount of salt in each tank. Hint: It is okay to use piecewise functions.

2. Suppose a disease spreading through a school can be modeled with the SIR model. How many of the susceptible children should be vaccinated in order to avoid an epidemic?

3. Suppose a disease is spreading through a school, with $S + I + R$ constant (so that $S' + I' + R' = 0$). Suppose that the per capita rate at which susceptibles are infected is proportional to the number of infectives, the per capita rate at which infectives recover is constant, and recovereds become susceptible again at a constant per capita rate. Draw a compartment diagram and construct the differential equation model.

4. Show that the differential equation

$$\frac{dx}{dt} = bx - dx^2$$

   can be written as

$$\frac{dx}{dt} = rx\left(1 - \frac{x}{K}\right),$$

   where $r = b$ and $K = b/d$.

5. Prove that there is no value of C for which the formula

$$x(t) = \frac{Ce^{rt}}{1 + Ce^{rt}/K}$$

   satisfies the initial condition $x(0) = K > 0$.

6. Solve equation 7.6 for C to obtain equation 7.7.

7. Show that equations (7.5) and (7.7) together yield the particular solution (7.8). Be careful with the singular solution.

8. Consider the differential equation

$$\frac{dx}{dt} = t^2 x^2.$$

Note that this equation is separable and that there is one equilibrium: $x_e = 0$.

   a. Find the general solution using the method of separation of variables.

   b. Find the particular solution given the initial condition $x(0) = 1/2$.

   c. Find the particular solution given the initial condition $x(0) = 0$.

9. Consider the differential equation

$$\frac{dx}{dt} = t\left(1 - x^2\right).$$

   a. Find all equilibria x_e.

   b. Find the general solution using the method of separation of variables. Hint: Write $(1 - x^2)$ as $(1 - x)(1 + x)$.

   c. Find the particular solution given the initial condition $x(0) = 2$.

   d. Find the particular solution given the initial condition $x(0) = -1$.

   e. Find the particular solution given the initial condition $x(0) = 1$.

10. The discrete-time logistic map

$$x_{t+1} = r x_t \left(1 - x_t\right),$$

considered in Exercise 9 of Chapter 3, is so-called because its quadratic nonlinearity reminds us of the well-known continuous-time logistic model discussed in this chapter

$$dx/dt = rx\,(1 - x/K)\,.$$

However, the discrete-time logistic map is *not* a good discrete analogue of the continuous-time logistic model. In fact, it turns out that the Beverton-Holt map

$$x_{t+1} = \frac{bx_t}{1 + cx_t},$$

considered in Exercise 8 of Chapter 3, is the discrete-time analogue of the continuous-time logistic model. In this exercise we will see why.

a. In the Beverton-Holt map, let $K = (b-1)/c$. Prove that the closed-form particular solution of the Beverton-Holt map with initial condition $x(0) = x_0$ is

$$x_t = \frac{Kx_0}{x_0 + (K - x_0)\,b^{-t}}. \tag{7.9}$$

Hint: Prove that the proposed solution satisfies the Beverton-Holt iterative map.

b. How do solutions (7.9) of the discrete-time Beverton-Holt map compare to solutions (7.8) of the continuous-time logistic differential equation?

BIBLIOGRAPHY

Kermack, W. O. and McKendrick, A. G. 1927. A contribution to the mathematical theory of epidemics. *Proceedings of the Royal Society of London A.* 115:700–721. DOI: 10.1098/rspa.1927.0118. [Historic paper introducing the SIR model with random infectious encounters, similar to the law of mass action in chemistry. Many models in epidemiology are variations of the SIR model.]

Pearl, R. and Reed, L. J. 1920. On the rate of growth of the population of the United States since 1790 and its mathematical representation. *Proceedings of the National Academy of Sciences of the United States of America* 6:275–288. DOI:10.1073/pnas.6.6.275. [Pearl and Reed rediscovered the Verhulst's logistic growth model, sometimes called the Verhulst-Pearl model.]

Verhulst, P.-F. 1838. Notice sur la loi que la population suit dans son accroissement. *Correspondance mathematique et physique* 10:113–121. [(Note on the law of population increase.) Verhulst introduces the logistic differential equation model of population growth, which is a modification of the Malthusian exponential growth model by the inclusion of crowding effects.]

Verhulst, P.-F. 1845. Recherches mathematiques sur la loi d'accroissement de la population. *Nouveaux Memoires de l'Academie Royale des Sciences et Belles-Lettres de Bruxelles* 18:1–42. [(Mathematical researches into the law of population growth increase.) Verhulst names the solution the "logistic curve."]

CHAPTER 8

Scalar Differential Equations

8.1 WHAT YOU SHOULD KNOW ABOUT THIS CHAPTER

A scalar differential equation is a differential equation with one state variable. In this chapter, we consider only *autonomous* differential equations (ones which do not explicitly depend on time). Instead of focusing on solution techniques, we focus on equilibria, stability, and bifurcation techniques. It is important for biologists to realize that, in general, it is not necessary to actually solve a nonlinear differential equation in order to study its equilibria and stability. The main tools come from first-semester calculus.

For the bifurcation approach to scalar differential equations given here and the section on human population growth, the authors are indebted to J. M. Cushing's textbook *Differential Equations: An Applied Approach* (Cushing 2004), from which co-author Henson spent many pleasant years teaching differential equations.

8.2 LINEAR EQUATIONS

We first consider the linear equation

$$\frac{dx}{dt} = rx \qquad (8.1)$$
$$x(0) = x_0,$$

where r is a real parameter. What are the equilibria of this model? The system is at equilibrium if and only if the derivative dx/dt is zero for all time t. Thus, the equilibrium equation (or fixed point equation) is

$$0 = rx.$$

To find the equilibria, we must solve the equilibrium equation for constant x. The extinction state $x_e = 0$ is the only equilibrium if $r \neq 0$. If $r = 0$, then every constant number x is an equilibrium.

The *general solution* (i.e., the set of all solutions) of the differential equation (8.1) is

$$x_g(t) = ce^{rt},$$

where c is any constant (Exercise 1). Now apply the initial condition $x(0) = x_0$ to obtain

$$x_0 = ce^{r0},$$

which implies $c = x_0$. Thus, the *particular solution* (i.e., the solution that satisfies the initial condition) is

$$x_p(t) = x_0 e^{rt}.$$

From this solution formula, we can see that if $r > 0$, then the extinction equilibrium $x_e = 0$ is unstable, because e^{rt} is growing over time. If $r < 0$, then the extinction state is asymptotically stable, because e^{rt} is approaching zero over time. If $r = 0$, then every real number x_0 is an equilibrium, and these equilibria are all neutrally stable. The bifurcation diagram in Figure 8.1 summarizes these dynamics as a function of the bifurcation parameter r. The bifurcation at $r = 0$ is called a *vertical bifurcation* (Figure 8.1).

For a specific example, consider the initial value problem

$$
\begin{aligned}
x' &= 2x \\
x(0) &= 5.
\end{aligned}
$$

The general solution is $x = ce^{2t}$. Plugging in the initial condition to obtain $c = 5$, we find the particular solution

$$x(t) = 5e^{2t}.$$

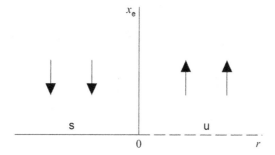

Figure 8.1: Bifurcation diagram for linear scalar differential equation. A vertical bifurcation occurs at $r = 0$.

8.2.1 Malthusian Growth

Now consider a population with constant per capita birth rate $b > 0$ and constant per capita death rate $d > 0$ and no emigration or immigration. The total influx (inflow rate) is bx, and the total outflux (outflow rate) is dx. Thus, we have

$$\frac{dx}{dt} = bx - dx = (b - d)\,x \tag{8.2}$$
$$x(0) = x_0.$$

The general solution is $x = ce^{(b-d)t}$. Applying the initial condition yields the particular solution

$$x(t) = x_0 e^{(b-d)t}.$$

Figure 8.2 gives time series graphs for the three cases $b < d$, $b = d$, and $b > d$. If $b < d$, then all nonzero solutions are decaying exponentially toward zero. If $b = d$, then the initial condition itself is an equilibrium solution. If $b > d$, then all nonzero solutions are growing exponentially.

The exponential growth when $b > d$ is called *Malthusian growth*, after Thomas Malthus, an economist and English curate who wrote the influential piece *An Essay on the Principle of Population* in 1798 (Malthus 1798).

Note that *linear models have exponential solutions.*

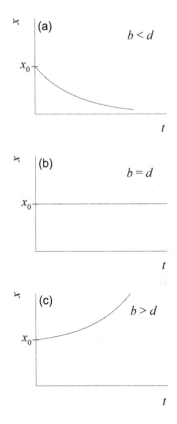

Figure 8.2: Time series for model (8.2). (a) $b < d$, (b) $b = d$, (c) $b > d$.

8.3 NONLINEAR EQUATIONS

8.3.1 Logistic Growth

Consider again equation (8.2). When b and d are assumed to be constant and $b > d$, the model exhibits Malthusian growth, which of course cannot be sustained indefinitely because of crowding effects. Sooner or later, the vital rates b and d will themselves become functions of population density. In this case, the population model becomes nonlinear and the population is said to have *density dependence*. For example, suppose that a population grows with constant per capita birth rate $b > 0$, but that the per capita death rate is proportional to the population size. Assume there is no emigration or immigration. The total population inflow rate is therefore bx, and the total population outflow rate has the form $(dx) x = dx^2$. Thus, we have the nonlinear model

$$\frac{dx}{dt} = bx - dx^2. \tag{8.3}$$

This is the famous logistic model, but it is traditionally written in a different form. Choose two new parameters r and K defined by $r = b$ and $K = b/d$. Then we can rewrite equation (8.3) as

$$\frac{dx}{dt} = rx \left(1 - \frac{x}{K}\right). \tag{8.4}$$

Equation (8.4) is the traditional form of the continuous-time *logistic model*. What are the dynamics? First let's compute the equilibria, that is, the constant solutions. The solution $x(t)$ is constant if and only if $dx/dt = 0$ for all time t, so the equilibrium equation is

$$rx \left(1 - \frac{x}{K}\right) = 0.$$

The equilibria are therefore $x_e = 0$ and $x_e = K$. Figure 8.3 illustrates the behavior of all possible solutions as a function of t. Note from Figure 8.3 that $x_e = 0$ is an unstable equilibrium, whereas $x_e = K$ is an asymptotically stable equilibrium.

The parameter r is called the *intrinsic growth rate*. It is the exponential rate of growth for small population sizes, when crowding effects are negligible. The parameter K is called the *carrying capacity of the environment*.

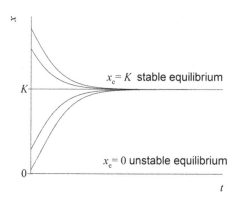

Figure 8.3: Illustration of all possible solutions of the logistic model (8.4).

8.3.2 Allee Effects

The *per capita growth rate* of a population of density $x(t)$ is defined to be

$$\frac{1}{x}\frac{dx}{dt}.$$

We tend to think of the cases in which density dependence causes the per capita growth rate to decrease if population density increases. Sometimes, however, it works the other way around so that an increase in population density x results in an increase in the per capita growth rate. This is called an *Allee effect*. Allee effects can happen when, for example, a population requires a threshold density in order to reproduce.

Consider again the logistic model

$$\frac{dx}{dt} = rx\left(1 - \frac{x}{K}\right),$$

where $r, K > 0$. The per capita growth rate is

$$r\left(1 - \frac{x}{K}\right).$$

Here, we see that an increase in x causes a decrease in the per capita growth rate (Figure 8.4). Thus, the logistic growth does not exhibit Allee effects.

We can modify the logistic equation to include an Allee effect:

$$\frac{dx}{dt} = rx\,(x - a)\left(1 - \frac{x}{K}\right), \tag{8.5}$$

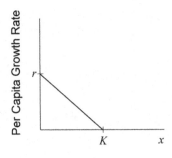

Figure 8.4: Per capita growth rate in the logistic model graphed against population size x.

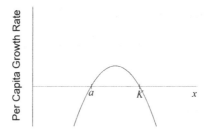

Figure 8.5: Per capita growth rate in model (8.5) graphed against population size x.

where $r, a, K > 0$ and $a < K$. The equilibria for this model are $x_e = 0, a, K$, and the per capita growth rate is

$$r\left(x - a\right)\left(1 - \frac{x}{K}\right).$$

In Figure 8.5, we see that at low population densities (small values of x), an increase in x causes an increase in the per capita growth rate. Thus, the dynamics have an Allee effect at low population sizes. Figure 8.6 illustrates the behavior of all possible solutions of model (8.5) in time t. Note from Figure 8.6 that $x_e = 0$ is asymptotically stable, $x_e = a$ is unstable, and $x_e = K$ is asymptotically stable. The equilibrium $x_e = a$ is called an *Allee threshold*. The initial population size must be above this threshold in order for the population to survive (Figure 8.6).

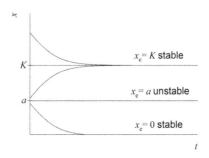

Figure 8.6: Illustration of all possible solutions of model (8.5).

8.3.3 The "Doomsday Model" of Human Population Growth

Consider again the Malthusian growth model

$$\frac{dx}{dt} = rx$$
$$x(0) = x_0 > 0$$

with $r > 0$. The particular solution is

$$x(t) = x_0 e^{rt}.$$

After a certain amount of time T, the population size will have doubled to $2x_0$. What is the doubling time T? That is, for what value of T is

$$2x_0 = x_0 e^{rT}?$$

Note that the initial population size x_0 cancels if it is nonzero, and we can solve for T to obtain a formula for the doubling time that does not depend on the initial population size:

$$T = \frac{\ln 2}{r}.$$

Note that *exponential growth is characterized by a constant doubling time.* Similarly, exponential decay is characterized by a constant half-life.

An interesting problem in the differential equations textbook by J. M. Cushing (Cushing 2004) took another look at human population growth using census data for years 1900–1990 from the United Nations (see bibliography entry for URL). In Table 8.1, we see that the doubling time was not constant, but was decreasing. A decreasing doubling time means that the human population was growing faster than any exponential function.

Consider the population model

$$\frac{dx}{dt} = rx^{\alpha} \tag{8.6}$$
$$x(0) = x_0$$

with constants $r, \alpha > 0$. For an exponentially growing population, $\alpha = 1$ and doubling times are constant. If $\alpha > 1$, the population is

TABLE 8.1   World Population Size 1900–1990

Year	World Population (billions)	Approximate Doubling Time (years)
1900	1.65	60.4
1910	1.75	50.8
1920	1.86	50.0
1930	2.07	40.6
1940	2.30	40.2
1950	2.52	30.7
1960	3.02	
1970	3.70	
1980	4.45	
1990	5.30	

growing faster than exponential and doubling times are *decreasing*. If $0 < \alpha < 1$, the population is growing, but it is growing more slowly than exponential and doubling times are *increasing*. The decreasing doubling times in Table 8.1 suggest that $\alpha > 1$ for the years shown.

To solve equation (8.6), we separate the variables as illustrated in Chapter 7:

$$\frac{dx}{x^\alpha} = rdt$$

$$\int \frac{dx}{x^\alpha} = \int rdt + C$$

$$\frac{x^{1-\alpha}}{1-\alpha} = rt + C \tag{8.7}$$

as long as $\alpha \neq 1$. Equation (8.7) is the general solution in implicit form. To find the particular solution, we apply the initial condition $x(0) = 1.65$, in which year 1900 is taken to be $t = 0$. This yields

$$C = \frac{1.65^{1-\alpha}}{1-\alpha}. \tag{8.8}$$

From equations (8.7) and (8.8), we see that the particular solution satisfies

$$x^{1-\alpha} = rt(1-\alpha) + 1.65^{1-\alpha},$$

and can be written

$$x_p(t) = \left[rt(1-\alpha) + 1.65^{1-\alpha} \right]^{1/(1-\alpha)}. \tag{8.9}$$

We need to estimate two parameters, r and α. In his textbook (Cushing 2004), Cushing accomplished this by requiring that the particular solution pass through the data points for years 1940 and 1990 ($t = 40$ and $t = 90$):

$$x(40) = 2.30$$
$$x(90) = 5.30.$$

Thus, we need to solve the system of two equations

$$2.30 = \left[40r\left(1-\alpha\right)+1.65^{1-\alpha}\right]^{1/(1-\alpha)} \tag{8.10}$$
$$5.30 = \left[90r\left(1-\alpha\right)+1.65^{1-\alpha}\right]^{1/(1-\alpha)}$$

for the two unknowns r and α.

The parameter estimates, here obtained from a computer, are (Exercise 2)

$$r = 3.6022 \times 10^{-3} \tag{8.11}$$
$$\alpha = 2.2633.$$

Substituting the values of α and r given in equation (8.11) into equation (8.9), we find

$$x\left(t\right) = \frac{1}{[(-4.551 \times 10^{-3})\,t + 0.5312]^{0.7916}}.$$

The strange and rather shocking fact from the point of view of modeling is that this faster-than-exponentially-growing solution *blows up in finite time.* That is, the solution *has a vertical asymptote* when

$$\left(-4.551 \times 10^{-3}\right)t + 0.5312 = 0,$$

that is, when $t = 116.7$. This time corresponds to the year 1900 + 116.7 = 2016.7, or about the year 2017. A similar model published in the journal *Science* in 1960, called the "Doomsday Model," predicted a doomsday of November 13, 2026 (von Foerster, Mora, and Amiot 1960).

The Doomsday Model and these various dates for "doomsday" are humorous, disturbing, and, for the modeling student, enlightening.

TABLE 8.2   World Population Size
2000–2020

Year	World Population (billions)
2000	6.14
2010	6.96
2020	7.79

They are humorous because it is obvious that, on a bounded planet, a population size cannot blow up in finite time. And we laugh because, at this writing, the year 2017 has already passed. Even so, this exercise is disturbing because one realizes that the dynamic mechanisms behind supra-exponential growth could deplete the earth's resources extremely quickly, leading to untold suffering. For the modeling student, the Doomsday Model makes a very important point about the interpretation of predictions by mathematical models. *When modeling dynamical systems, it is important to remember that model predictions are based on model assumptions.* Predictions change if underlying assumptions are altered.

Let's now consider the United Nations data for years 2000–2020 (Table 8.2) (see bibliography for URL).

The data for 2000–2020 show almost linear growth, which is sub-exponential. Indeed, for model (8.6), applied to the data in Table 8.2, α would be close to zero. Clearly, the dynamic mechanisms changed, and a model for the data during 1900–2020 should be more like the logistic model. At this point, our concern turns to the predicted value of the carrying capacity K (Exercise 3). When world population levels off, will there be enough resources for every human being to have a decent standard of living while preserving the rest of the biodiversity and beauty of the planet?

Scientists have a duty to make it clear to the public that *model predictions are always conditional projections.* Predictions change if underlying model assumptions are altered. As the ghost said about Tiny Tim in Charles Dickens' novel *A Christmas Carol*, "If these shadows remain unaltered by the Future, none other of my race...will find him here."

8.4   LINEARIZATION

Consider the differential equation

$$\frac{dx}{dt} = x\,(x-1)\,(x-2).$$

Let $f(x) = x\,(x-1)\,(x-2)$. The equilibria are

$$x_e = 0, 1, 2.$$

Figure 8.7a shows the graph of $f(x)$ vs. x. The roots of f are the equilibria. When the graph of f is above the x-axis, $dx/dt > 0$ and so $x(t)$ is increasing. We denote this with a time arrow to the right on the x-axis. When the graph of f is below the x-axis, we insert an arrow to the left on the x-axis. If we extract the x-axis by itself with the equilibria shown as circles and with the time arrows in each interval between equilibria, we obtain the *phase line portrait* (Figure 8.7b). If the arrows point toward the equilibrium on each side, then the equilibrium is a *sink* and is asymptotically stable. If the arrows on each side point away from the equilibrium, then the equilibrium is a *source* and is unstable. All possible time series for x vs. t are shown in Figure 8.7c.

Note from the geometry in Figure 8.7a that if the graph of f is increasing through the equilibrium as a function of x, then the arrows point away from the equilibrium, and the equilibrium is unstable. Similarly, if f is decreasing through the equilibrium as a function of x, then the arrows point toward the equilibrium, which is then asymptotically stable. In the current example, we have

$$\frac{df}{dx}(0) \;>\; 0 \Longrightarrow x_e = 0 \text{ unstable}$$

$$\frac{df}{dx}(1) \;<\; 0 \Longrightarrow x_e = 1 \text{ asymptotically stable}$$

$$\frac{df}{dx}(2) \;>\; 0 \Longrightarrow x_e = 2 \text{ unstable.}$$

Note: Be careful! df/dx is NOT the same as dx/dt.

These considerations lead to the following definition and theorem.

(a)

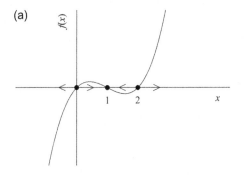

(b)

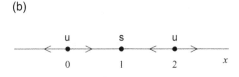

(c)

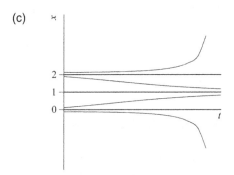

Figure 8.7: (a) Graph of $f(x)$ vs. x. The roots of f are the equilibria. (b) Phase line portrait. (c) Time series for $x(t)$ vs. t showing equilibrium solutions (horizontal lines) and illustrating the behavior of all other solutions.

Definition 8.1 *Consider the differential equation*

$$\frac{dx}{dt} = f(x).$$

An equilibrium x_e is **hyperbolic** *if and only if* $\frac{df}{dx}(x_e) \neq 0$.

Theorem 8.1 *(Linearization Theorem) Let x_e be a hyperbolic equilibrium of*

$$\frac{dx}{dt} = f(x)$$

and suppose that the function f is continuously differentiable in x (that is, df/dx exists and is continuous). Define

$$\lambda = \frac{df}{dx}(x_e).$$

Then

$$\lambda > 0 \Longrightarrow x_e \text{ is unstable}$$
$$\lambda < 0 \Longrightarrow x_e \text{ is asymptotically stable.}$$

As you will see in the next chapter, the number λ is a very important number called the *eigenvalue*. Figure 8.8 summarizes the relationships between sinks, sources, and eigenvalues. *Note that for $\lambda = 0$ (the nonhyperbolic case) you cannot conclude anything from linearization; the linearization theorem does not apply. See Figure 8.9.*

For an example, consider the logistic model

$$\frac{dx}{dt} = rx\left(1 - \frac{x}{K}\right),$$

where $r, K > 0$. The equilibria are $x_e = 0, K$. Also, $f(x) = rx(1 - x/K)$. It is easiest to use the product rule to find $df/dx =$

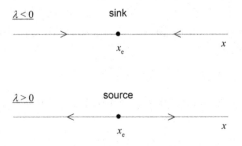

Figure 8.8: Phase line portraits for hyperbolic equilibria. The linearization theorem applies, and the sign of the eigenvalue λ determines stability.

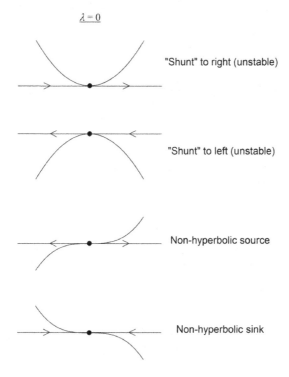

$\lambda = 0$

"Shunt" to right (unstable)

"Shunt" to left (unstable)

Non-hyperbolic source

Non-hyperbolic sink

Figure 8.9: Phase line portrait possibilities for nonhyperbolic equilibria. The linearization theorem does not apply.

$r\left(1 - x/K\right) - rx/K$, because as you plug in each equilibrium, all terms will zero out except one. At $x_e = 0$, $df/dx = r > 0$, so the equilibrium $x_e = 0$ is unstable. At $x_e = K$, $df/dx = -r < 0$, so the equilibrium $x_e = K$ is asymptotically stable (Figure 8.10).

For another example, let's consider

$$\frac{dx}{dt} = x^2 \left(x - a\right),$$

where $a \neq 0$. The equilibria are $x_e = 0, a$. Also, $f\left(x\right) = x^2 \left(x - a\right)$, so, using the product rule, we have $df/dx = 2x\left(x - a\right) + x^2$. At $x_e = 0$, $df/dx = 0$, and at $x_e = a$, $df/dx = a^2$. Thus, from the linearization theorem, the equilibrium $x_e = a$ is unstable and no conclusion can be drawn about $x_e = 0$. The geometric method (Figure 8.11) shows that $x_e = 0$ is a shunt (and thus unstable).

Figure 8.10: Phase line portrait for the logistic model.

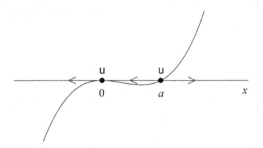

Figure 8.11: Phase line portrait for $dx/dt = x^2 (x - a)$, shown for $a > 0$. For $a < 0$, $x_e = 0$ becomes a shunt to the right.

For a final example, let's consider

$$\frac{dx}{dt} = x (x - a) (10 - x),$$

where $a \in R$. The equilibria are $x_e = 0, a, 10$. Using the extended product rule, we find that $df/dx = (x - a) (10 - x) + x (10 - x) - x (x - a)$. At $x_e = 0$, $df/dx = -10a$. At $x_e = a$, $df/dx = a (10 - a)$. At $x_e = 10$, $df/dx = -10 (10 - a)$. The stabilities of the equilibria depend on the value of a. For example, for $x_e = 0$ we have $df/dx > 0$ for $a < 0$ and $df/dx < 0$ for $a > 0$. Thus, the equilibrium $x_e = 0$ is unstable for $a < 0$ and asymptotically stable for $a > 0$. Similarly, the equilibrium $x_e = a$ is unstable for $0 < a < 10$ and asymptotically stable for $a < 0$ and $a > 10$. The equilibrium $x_e = 10$ is asymptotically stable for $a < 10$ and unstable for $a > 10$. The phase line portraits and bifurcation diagram in Figures 8.12–8.13 summarize this information.

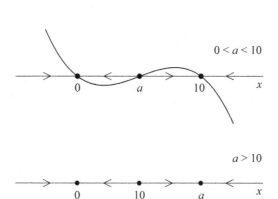

Figure 8.12: Possible phase line portraits for dx/dt = $x(x-a)(10-x)$.

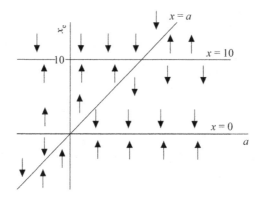

Figure 8.13: Bifurcation diagram for $dx/dt = x(x-a)(10-x)$, using a as the bifurcation parameter.

8.5 BIFURCATIONS

In this section, we illustrate some canonical bifurcations for scalar differential equations.

8.5.1 Transcritical Bifurcation

Consider the equation

$$x' = x\,(x - a),$$

where the "prime" denotes the time derivative dx/dt and $a \in R$ is a bifurcation parameter. The equilibria are $x_e = 0$ and $x_e = a$. In the bifurcation diagram, these two equilibrium branches cross at $a = 0$ (Figure 8.14) at a *transcritical bifurcation*. In Exercise 6, you will prove that for $a < 0$, the equilibrium $x_e = 0$ is unstable and $x_e = a$ is asymptotically stable, and for $a > 0$, the equilibrium $x_e = 0$ is asymptotically stable and $x_e = a$ is unstable. This means that a typical *exchange of stability* occurs at the bifurcation point $a = 0$.

The correct interpretation of Figure 8.14 is important. For each fixed value of the parameter a, one obtains a different differential equation. The phase line portrait for that differential equation is the vertical line through that value of a. We might say the dynamics for any fixed a "live" on a vertical line. Thus, the bifurcation diagram is the collection of all possible phase line portraits, slotted together as vertical lines, as a function of parameter a. In Exercise 6, you will determine the stabilities in two ways, first by using the linearization theorem, and then by using the geometric method. The geometric

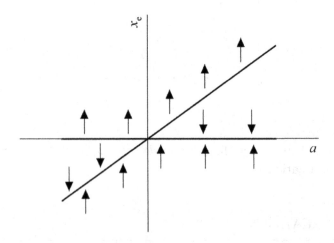

Figure 8.14: Transcritical bifurcation at $a = 0$ with typical exchange of stability.

method consists of drawing all possible phase line portraits (as a function of a) with their stability arrows and then fitting them together vertically to determine the stabilities of the equilibrium branches in the bifurcation diagram.

8.5.2 Saddle-Node Bifurcation

Consider

$$x' = x^2 - a.$$

The equilibria are $x_e = \pm\sqrt{a}$. In the bifurcation diagram, this is a single branch of equilibria (Figure 8.15) with a *saddle-node bifurcation* at $a = 0$. For $a < 0$, there is no equilibrium. In Exercise 7, you will show that for $a > 0$, the upper part of the branch $x_e = \sqrt{a}$ is unstable and the lower part $x_e = -\sqrt{a}$ is asymptotically stable.

8.5.3 Pitchfork Bifurcation

Now consider the equation

$$x' = x\left(x^2 - a\right).$$

The equilibria are $x_e = 0$ and $x_e = \pm\sqrt{a}$. In the bifurcation diagram, these two branches cross at $a = 0$ in a *pitchfork bifurcation* (Figure 8.16). In Exercise 8, you will show that for $a < 0$, the equilibrium $x_e = 0$ is unstable, and for $a > 0$, the equilibrium $x_e = 0$ is asymptotically

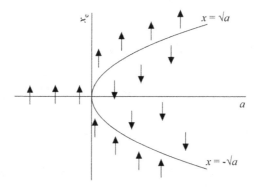

Figure 8.15: Saddle-node bifurcation at $a = 0$.

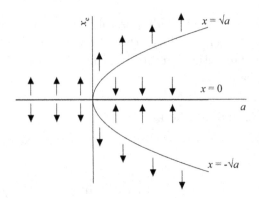

Figure 8.16: Pitchfork bifurcation at $a = 0$.

stable and $x_e = \pm\sqrt{a}$ are unstable. An exchange of stability occurs at the bifurcation point $a = 0$.

8.5.4 Hysteresis

Finally, consider

$$x' = x^3 - x + a.$$

It is not so easy here to find the equilibria, but we note that the bifurcation parameter a in $f(x) = x^3 - x + a$ vertically translates the graph of f. First, graph $f(x)$ for $a = 0$. The roots are $-1, 0$, and 1 (Figure 8.17). From the geometry we can include stability arrows

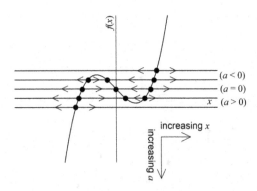

Figure 8.17: $f(x)$ vs. x for various values of the parameter a. The graph is shifted vertically upward when $a > 0$, which corresponds to dropping the horizontal axis, etc.

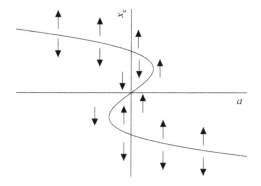

Figure 8.18: Hysteresis.

on the x-axis. If we "tune" a to small positive numbers, we raise the graph of f slightly. The lower equilibrium moves to the left, and the two upper equilibrium move toward each other. As we continue to increase the parameter a, the two upper equilibria collide and wink out of existence, while the lower equilibrium moves off toward $-\infty$. If we tune a from zero down to negative values, a similar thing happens. The original graph of $f(x)$ drops, the lower two equilibria move together and annihilate each other, while the upper equilibrium moves off to $+\infty$ (Figure 8.17).

If we graph the path of the equilibria in Figure 8.17 as a function of a, we obtain the bifurcation diagram in Figure 8.18. The stability arrows in Figure 8.18 come from those in Figure 8.17 via the geometric method. This bifurcation diagram is called a *hysteresis* and is composed of two saddle-node bifurcations. In Exercise 9, you will use the linearization theorem to verify the stabilities shown in Figure 8.18. Exercise 10 illustrates a *hysteresis loop*, which occurs if the stabilities of the segments of the branch are reversed from what they are in Figure 8.18.

8.6 EXERCISES

1. Use the method of separation of variables from Chapter 7 to show that the general solution of $dx/dt = rx$ is $x(t) = ce^{rt}$ for $c \in R$. Keep track of lost equilibrium solutions and the values of constants as we did in Chapter 7.

2. Use the computer to estimate the parameters in equation (8.11) from the system of equations (8.10). Hint: To numerically solve $g(x) = h(x)$, one can use the Nelder-Mead algorithm to minimize the function $(g(x) - h(x))^2$; if the minimum is zero, you have obtained a solution. Don't forget that you may need to try different initial guesses for the downhill method.

3. In this exercise, you will fit the logistic model

$$\frac{dx}{dt} = rx \left(1 - \frac{x}{K}\right)$$

to the world population data for the years 1900–2020 in Tables 8.1–8.2 in order to estimate the value at which the population is expected to level off, that is, the value of the carrying capacity K. One could approach this in several different ways, but here we specify two such methods.

a. Try fitting the particular solution

$$x(t) = \frac{Kx_0}{x_0 + (K - x_0)\, e^{-rt}}$$

of the logistic model (which we obtained in Chapter 7) to the data using the method of nonlinear least squares under the assumption of demographic noise. Here, $x_0 = 1.65$, where the units are in billions and $t = 0$ denotes year 1900. Hint: You will need to use your skills from Chapter 2 to write a program that minimizes the RSS and estimates the values of r and K (the latter of which is in units of billions). You should be able to modify one of your existing programs from that chapter. Remember to try multiple starting values for your parameters.

b. Now try fitting the general solution

$$x(t) = \frac{Ce^{rt}}{1 + Ce^{rt}/K}$$

of the logistic model (which we obtained in Chapter 7) to the data using the method of nonlinear least squares under the assumption of demographic noise. Here, you are letting the initial condition "float," meaning that you are not forcing the curve to fit the data point at $t = 0$, and you are estimating C as well as r and K.

c. Graph the scatter plot of data along with the fitted logistic curves for both fitting methods above.

d. What are your thoughts about the logistic model as a description of world population growth during the years 1900–2020? Can you propose a simple model that you think might fit better?

e. What are your thoughts about world population growth, and have they changed in any way after reading this chapter?

4. In this problem, you will do a complete equilibrium stability analysis for the logistic model with Allee effects

$$\frac{dx}{dt} = rx(x - a)\left(1 - \frac{x}{K}\right)$$
$$r, K > 0.$$

a. Find all the equilibria.

b. In the context of the linearization theorem, what is $f(x)$ in this model? What is df/dx? Use the extended product rule when computing df/dx.

c. Use the linearization theorem to determine conditions on the parameter a that guarantee the stability/instability of each equilibrium.

d. Assume $r, K > 0$ are fixed, and assume you can manipulate the parameter a. On one graph, plot each equilibrium x_e versus a. Use the ranges $-\infty < x < \infty$ and $-\infty < a < \infty$. This is a bifurcation diagram.

e. Indicate stabilities on your bifurcation diagram.

5. Consider the nonlinear differential equation

$$\frac{dx}{dt} = f(x) \qquad (8.12)$$

with hyperbolic equilibrium x_e. Recall that x_e is an equilibrium if and only if $f(x_e) = 0$. Before beginning this problem, you should review Section 3.5 of Chapter 3.

a. In calculus, you learned how to linearize a function about a point by means of a tangent line. Use that procedure to prove that

$$f(x) \approx \lambda(x - x_e)$$

for $x \approx x_e$, where $\lambda = \frac{df}{dx}(x_e)$. Explain why $\lambda \neq 0$ in this particular problem.

b. Write down a linear differential equation that approximates the nonlinear one (8.12) for $x \approx x_e$.

c. Define the variation or displacement of the system from equilibrium to be $u(t) = x(t) - x_e$. How is du/dt related to dx/dt?

d. If the system is near equilibrium, that is, if $x \approx x_e$, then what can we say about the value of the variation u?

e. Use the change of variables $u = x - x_e$ to derive the differential equation

$$\frac{du}{dt} = \lambda u. \qquad (8.13)$$

This is the *variation equation*. It is also called the *linearization of $dx/dt = f(x)$ at the equilibrium x_e*.

f. The equilibrium of equation (8.12) is x_e. The equilibrium of equation (8.13) is $u = 0$. Explain why this makes sense intuitively in terms of the meaning of $u(t)$.

g. What is the general solution of equation (8.13)?

h. Using the general solution of equation (8.13), explain why

$$\lambda > 0 \Longrightarrow u_e = 0 \text{ unstable}$$
$$\lambda < 0 \Longrightarrow u_e = 0 \text{ asymptotically stable.}$$

i. Intuitively, how should the sign of λ and the resulting stability/instability of $u_e = 0$ be related to the stability/instability of x_e? Explain. Compare your answer to the linearization theorem.

j. How is the linearization theorem for one-dimensional *differential equations* different from the linearization theorem for one-dimensional *difference equations*?

6. In this problem, you will verify the stabilities of the equilibrium branches in Figure 8.14.

   a. Use the linearization theorem to verify the stabilities of the equilibrium branches in Figure 8.14.

   b. Use the geometric method to verify the stabilities of the equilibrium branches in Figure 8.14. The geometric method consists of drawing all possible phase line portraits (as a function of a) with their stability arrows and then fitting them together vertically to determine the stabilities of the equilibrium branches in the bifurcation diagram.

7. Repeat Exercise 6 for Figure 8.15.

8. Repeat Exercise 6 for Figure 8.16.

9. Use the linearization theorem to verify the stabilities of the equilibrium branches in Figure 8.18.

10. Consider

$$x' = -x^3 + x + a.$$

   a. Draw the bifurcation diagram, using a as a bifurcation parameter. Include stability arrows.

   b. The bifurcation diagram contains a hysteresis composed of two saddle-node bifurcations. Prove that the saddle-node bifurcations occur at $a = \pm 2/(3\sqrt{3})$.

   c. Consider your bifurcation diagram. Imagine "tuning" the parameter a from $-\infty$ to $+\infty$. For each fixed value of a, we obtain a specific differential equation, and the system's phase line portrait is the vertical line through that value of a. We might say that the dynamics always "live" on a vertical line. If a is fixed with $a < -2/(3\sqrt{3})$, the system approaches an equilibrium on the lower branch of equilibria, which is stable. As a is tuned through the bifurcation value $a = 2/(3\sqrt{3})$ and becomes slightly larger than $2/(3\sqrt{3})$, the lower branch disappears and the system must approach an equilibrium on the upper branch, which is also stable. If a is then tuned back down through the bifurcation value

$a = -2/(3\sqrt{3})$, which branch does the system approach? Using two large vertical arrows, draw the hysteresis loop on the bifurcation diagram. This is the loop traced out by the system's final states as the parameter a is increased through $a = 2/(3\sqrt{3})$ and then decreased through $a = -2/(3\sqrt{3})$.

11. This exercise uses the skills developed in Exercise 10. Consider the logistic model for a fishery with an extra removal rate due to fishing. Suppose the rate at which fish are removed due to fishing is a constant h fish per unit time. The modified model is then

$$\frac{dx}{dt} = rx\left(1 - \frac{x}{K}\right) - h$$

with $r, K > 0$ and $h \geq 0$.

a. Draw the bifurcation diagram using h as the bifurcation parameter. Include stability arrows. Hint: The parameter h translates the graph of f vertically.

b. Find the value of h for which the system has a saddle-node bifurcation.

c. What happens in this fishery if h is increased slightly beyond the bifurcation point? This unheralded collapse is sometimes called a *blue-sky bifurcation*, because it comes "out of the blue."

d. If fishing is then reduced to levels below the bifurcation value of h, does the fishery recover?

e. There is no hysteresis loop in this bifurcation diagram. What must happen for the fishery to recover?

12. Draw the bifurcation diagrams for the following systems, using r as a bifurcation parameter. Include stability arrows.

a. $dx/dt = (r - 2)(x - r)$
b. $dx/dt = (x - 2)(x - r)$
c. $dx/dt = \sin(r)$
d. $dx/dt = x(x - r)^2$
e. $dx/dt = (x - r^2)(r - x)$

BIBLIOGRAPHY

Cushing, J. M. 2004. *Differential Equations: An Applied Approach.* Prentice-Hall, Upper Saddle River, NJ. [A good source for more phase line portrait and bifurcation diagram exercises, as well as many interesting applied projects using differential equations, including the "Doomsday Model."]

Malthus, T. R. 1798. *An Essay on the Principle of Population.* J. Johnson, London. [Malthus noted that there are not enough resources to keep up with geometric (that is, exponential) population growth.]

United Nations Population Division. https://population.un.org.

von Foerster, H., Mora, P. M., and Amiot, L. W. 1960. Doomsday: Friday, 13 November, A.D. 2026. *Science* 132:1291–1295. DOI: 10.1126/science.132.3436.1291. [As the authors wrote, "At this date human population will approach infinity if it grows as it has grown in the last two millenia."]

CHAPTER 9

Systems of Differential Equations

9.1 WHAT YOU SHOULD KNOW ABOUT THIS CHAPTER

In this chapter, we consider how to model continuous-time systems whose descriptions require more than one state variable. We consider only *autonomous* systems of differential equations (ones which do not explicitly depend on time). Before beginning this chapter, you should review basic matrix operations and introductory linear algebra by working through the self-guided tutorial in Appendix A. A course in linear algebra is not a prerequisite for this textbook; everything you need to know from linear algebra for this and other chapters is contained in Appendix A. You will need a computer program to quickly graph phase plane portraits. The free Java version of PPLANE (Castellanos and Polking) is very easy to use for this purpose.

In this chapter, you will need to know how to take partial derivatives, a topic from third-semester calculus that we already encountered in Chapter 5. As a reminder, they are easy to learn. If you have a function of, say, two variables $f(x, y)$, then the partial derivative of f with respect to x, denoted $\partial f/\partial x$, is the derivative of f with respect to x while holding y constant. So, for example, if $f(x, y) = x^2 + y^2 + x^3 y^3$, then the partial derivatives are (showing zeros to make a point):

$$\frac{\partial f}{\partial x} = 2x + 0 + y^3 \left(3x^2\right)$$

DOI: 10.1201/9781003265382-12

$$\frac{\partial f}{\partial y} = 0 + 2y + x^3 \left(3y^2\right).$$

If this is your first experience with partial derivatives, work a few problems in a calculus book to get comfortable with them.

The main ideas in this chapter are to learn how to solve and analyze linear systems and then to apply those techniques to nonlinear systems by means of linearization. You should keep track of how these methods parallel those for discrete-time systems in Chapter 5. Thus, before beginning this chapter you should reread the sections in Chapter 5 that deal with solving linear discrete-time systems and the linearization of nonlinear discrete-time systems.

9.2   LINEAR SYSTEMS OF ODES AND PHASE PLANE ANALYSIS

Consider the linear system of ordinary differential equations (ODEs)

$$\begin{aligned} x' &= ax + by \\ y' &= cx + dy, \end{aligned} \tag{9.1}$$

in which the "prime" indicates the derivative with respect to time t. System (9.1) is linear because the terms are linear functions of the state variables x and y and their derivatives. The equilibria are the solutions of the algebraic system

$$\begin{aligned} 0 &= ax + by \\ 0 &= cx + dy. \end{aligned}$$

The only equilibrium is $(0,0)$ as long as the determinant of the coefficient matrix is nonzero, that is, as long as

$$\begin{vmatrix} a & b \\ c & d \end{vmatrix} \neq 0.$$

In what follows, we will always assume this holds.

There are six generic types of linear two-dimensional systems ("generic" meaning those with distinct, nonzero eigenvalues). We use an example to illustrate each type. Carefully work through the details of the first example; the remaining five examples are presented without showing all the work. Make sure you know how to work out the details of each example by hand.

9.2.1 Unstable Node

Consider

$$x' = 2x + 2y \tag{9.2}$$
$$y' = \frac{1}{2}x + 2y.$$

To solve this system of differential equations, we propose an Ansatz. Remembering from previous chapters that the linear scalar differential equation $x' = rx$ has exponential solutions of the form $x = ce^{rt}$, we propose an Ansatz of the form

$$x(t) = v_1 e^{\lambda t}$$
$$y(t) = v_2 e^{\lambda t}$$

and look for values of λ, v_1, and v_2, where v_1 and v_2 are not both zero. Plugging the Ansatz into the system of differential equations (9.2), we have

$$\lambda v_1 e^{\lambda t} = 2v_1 e^{\lambda t} + 2v_2 e^{\lambda t}$$
$$\lambda v_2 e^{\lambda t} = \frac{1}{2}v_1 e^{\lambda t} + 2v_2 e^{\lambda t},$$

or

$$(2 - \lambda) v_1 + 2v_2 = 0 \tag{9.3}$$
$$\frac{1}{2}v_1 + (2 - \lambda) v_2 = 0.$$

This algebraic system can be visualized as two lines in the v_1-v_2 plane, both passing through the origin since $(v_1, v_2) = (0, 0)$ is a solution. In order for two such lines to have a nontrivial solution (v_1, v_2), they must be coincident (i.e., the same line), which is true if and only if

$$\begin{vmatrix} 2 - \lambda & 2 \\ \frac{1}{2} & 2 - \lambda \end{vmatrix} = 0.$$

This holds if and only if

$$(2 - \lambda)^2 - 1 = 0 \tag{9.4}$$
$$(2 - \lambda)^2 = 1$$
$$2 - \lambda = \pm 1$$
$$\lambda = 1, 3.$$

Equation (9.4) is called the *characteristic equation*, and its solutions $\lambda = 1, 3$ are the *eigenvalues* of the system.

Now we consider each eigenvalue in turn and look for the corresponding v_1, v_2, not both zero. If $\lambda = 1$, then from system (9.3) we obtain

$$
\begin{aligned}
v_1 + 2v_2 &= 0 \\
\frac{1}{2}v_1 + v_2 &= 0.
\end{aligned}
$$

Thus, $v_1 = -2v_2$, so an eigenvector belonging to $\lambda = 1$ is $\begin{pmatrix} 2 \\ -1 \end{pmatrix}$. The corresponding eigensolution is

$$
\begin{pmatrix} x \\ y \end{pmatrix} = \begin{pmatrix} 2 \\ -1 \end{pmatrix} e^t.
$$

If $\lambda = 3$, then

$$
\begin{aligned}
-v_1 + 2v_2 &= 0 \\
\frac{1}{2}v_1 - v_2 &= 0.
\end{aligned}
$$

Thus, $v_1 = 2v_2$, so an eigenvector belonging to $\lambda = 3$ is $\begin{pmatrix} 2 \\ 1 \end{pmatrix}$ and the corresponding eigensolution is

$$
\begin{pmatrix} x \\ y \end{pmatrix} = \begin{pmatrix} 2 \\ 1 \end{pmatrix} e^{3t}.
$$

The general solution is a linear combination of the two independent eigensolutions:

$$
\begin{pmatrix} x \\ y \end{pmatrix} = c_1 \begin{pmatrix} 2 \\ -1 \end{pmatrix} e^t + c_2 \begin{pmatrix} 2 \\ 1 \end{pmatrix} e^{3t}.
$$

Here, you should think of the $c_1 e^t$ and $c_2 e^{3t}$ as scalar multipliers on the eigenvectors that stretch or shrink the eigenvectors.

In order to graph the general solution, that is, the set of all solutions, we graph temporal orbits in the *phase plane*, that is, *x-y* space. The temporal aspect is denoted by arrows indicating increasing time. First, we graph the orbits of the eigensolutions. These are simply the lines in the phase plane passing through the origin that are determined by the eigenvectors, with appropriate time arrows (Figure 9.1). These lines are called *manifolds*. Both of the manifolds in this example are called "unstable manifolds" because the system moves away from the origin as time progresses (the scalars e^t and e^{3t} are increasing). For other solutions, as $t \to \infty$, the scalar e^{3t} dominates the solution, which begins to roughly parallel the manifold determined by eigenvalue $(2,1)^\top$. On the other hand, when $t \to -\infty$, the scalar e^t begins to dominate, and, going backward in time, solutions roughly parallel the manifold determined by $(2,-1)^\top$. Figure 9.1 shows the phase plane portrait for this system with the manifolds and a few other orbits. This type of phase plane portrait for the equilibrium (here, the origin) is called an *unstable node*.

9.2.2 Asymptotically Stable Node

Consider the system

$$\begin{aligned} x' &= -3x - y \\ y' &= 2x. \end{aligned}$$

Following the same procedure as in the previous section, we find the eigenvalues by subtracting λ off the diagonal of the coefficient matrix

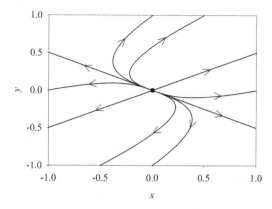

Figure 9.1: The equilibrium (0,0) is an unstable node.

and setting the determinant equal to zero,

$$\begin{vmatrix} -3 - \lambda & -1 \\ 2 & 0 - \lambda \end{vmatrix} = 0,$$

so that

$$\begin{aligned} -\lambda\left(-3 - \lambda\right) + 2 &= 0 \\ \lambda^2 + 3\lambda + 2 &= 0 \\ \left(\lambda + 1\right)\left(\lambda + 2\right) &= 0. \end{aligned}$$

The solutions of this characteristic equation are the eigenvalues $\lambda = -1, -2$. Using the procedure from the previous section to find the corresponding eigenvectors, we obtain

$$\begin{pmatrix} -1 \\ 1 \end{pmatrix} \text{ for } \lambda = -2 \text{ and } \begin{pmatrix} 1 \\ -2 \end{pmatrix} \text{ for } \lambda = -1.$$

Thus, the general solution is

$$\begin{pmatrix} x \\ y \end{pmatrix} = c_1 \begin{pmatrix} 1 \\ -2 \end{pmatrix} e^{-t} + c_2 \begin{pmatrix} -1 \\ 1 \end{pmatrix} e^{-2t}.$$

Here, both manifolds are stable (because the scalars e^{-t} and e^{-2t} are approaching zero). The term with e^{-t} dominates in forward time, and the term with e^{-2t} dominates in backward time, as shown in Figure 9.2. This phase plane portrait is an *asymptotically stable node* (Figure 9.2).

9.2.3 Saddle

Consider the system

$$\begin{aligned} x' &= x \\ y' &= -x - 2y. \end{aligned}$$

The eigenvalues are the solutions of

$$\begin{vmatrix} 1 - \lambda & 0 \\ -1 & -2 - \lambda \end{vmatrix} = 0,$$

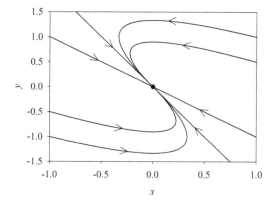

Figure 9.2: The equilibrium $(0,0)$ is an asymptotically stable node.

that is, the solutions of the characteristic equation

$$(1 - \lambda)(-2 - \lambda) = 0.$$

The eigenvalues are $\lambda = 1, -2$, and the corresponding eigenvectors are (make sure you can obtain these by hand)

$$\begin{pmatrix} 0 \\ 1 \end{pmatrix} \text{ for } \lambda = -2 \text{ and } \begin{pmatrix} -3 \\ 1 \end{pmatrix} \text{ for } \lambda = 1.$$

Thus, the general solution is

$$\begin{pmatrix} x \\ y \end{pmatrix} = c_1 \begin{pmatrix} 0 \\ 1 \end{pmatrix} e^{-2t} + c_2 \begin{pmatrix} -3 \\ 1 \end{pmatrix} e^t.$$

Here, we have a stable manifold and an unstable manifold, which gives rise to a *saddle* (Figure 9.3).

Note that a saddle is unstable. In two dimensions, it has one unstable direction and one stable direction. In higher dimensions, there is at least one unstable direction and at least one stable direction.

9.2.4 Center

Consider the harmonic oscillator

$$\begin{aligned} x' &= y \\ y' &= -x. \end{aligned} \tag{9.5}$$

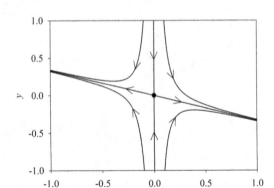

Figure 9.3: The equilibrium (0,0) is a saddle. Saddles are unstable.

The characteristic equation for system (9.5) is

$$\begin{vmatrix} -\lambda & 1 \\ -1 & -\lambda \end{vmatrix} = 0,$$

that is,

$$\lambda^2 + 1 = 0.$$

The eigenvalues $\lambda = \pm i$ are complex numbers, in this case purely imaginary numbers. For $\lambda = i$, we find the eigenvalues as before:

$$\begin{aligned} -iv_1 + v_2 &= 0 \\ -v_1 - iv_2 &= 0. \end{aligned}$$

If we take $v_1 = i$, then $v_2 = -1$. For $\lambda = -i$, we have

$$\begin{aligned} iv_1 + v_2 &= 0 \\ -v_1 + iv_2 &= 0. \end{aligned}$$

If we take $v_1 = i$, then $v_2 = 1$. Thus, we can express the general solution over the complex numbers as

$$\begin{pmatrix} x \\ y \end{pmatrix} = c_1 \begin{pmatrix} i \\ -1 \end{pmatrix} e^{it} + c_2 \begin{pmatrix} i \\ 1 \end{pmatrix} e^{-it}. \tag{9.6}$$

We would like to obtain from this a general solution over R^2.

From Euler's formula $e^{it} = \cos t + i \sin t$ and the facts that $\cos t$ is even and $\sin t$ is odd, one can obtain the complex solution (Exercise 1)

$$\begin{pmatrix} x \\ y \end{pmatrix} = (c_2 - c_1) \begin{pmatrix} \sin t \\ \cos t \end{pmatrix} + i (c_1 + c_2) \begin{pmatrix} \cos t \\ -\sin t \end{pmatrix}. \tag{9.7}$$

A theorem from differential equations says that the real and imaginary parts of this complex solution are two linearly independent real solutions, in this case

$$\begin{pmatrix} \sin t \\ \cos t \end{pmatrix} \text{ and } \begin{pmatrix} \cos t \\ -\sin t \end{pmatrix}.$$

The general solution over R^2 is therefore

$$\begin{pmatrix} x \\ y \end{pmatrix} = c_1 \begin{pmatrix} \sin t \\ \cos t \end{pmatrix} + c_2 \begin{pmatrix} \cos t \\ -\sin t \end{pmatrix}.$$

It is straightforward to check that the component-wise form

$$\begin{aligned} x &= c_1 \sin t + c_2 \cos t \\ y &= c_1 \cos t - c_2 \sin t \end{aligned} \tag{9.8}$$

satisfies system (9.5) and that the solutions are closed curves in the phase plane (Exercise 3). The direction of rotation can be determined by choosing a test point, say $x = 1$, $y = 0$, and checking the signs of the time derivatives at that point ($x' = 0$, $y' = -1$). This tells us that y is decreasing at that point, and so the rotation is clockwise in forward time.

When eigenvalues are purely imaginary, as in this example, the phase plane portrait is called a *center* (Figure 9.4). A center is neutrally stable.

9.2.5 Unstable Spiral

Consider the system

$$\begin{aligned} x' &= x - y \\ y' &= x + y. \end{aligned}$$

Check that the eigenvalues are $\lambda = 1 \pm i$. Note that $e^{(1 \pm i)t} = e^t e^{\pm it} = e^t (\cos t \pm i \sin t)$. As in the previous example, the system oscillates,

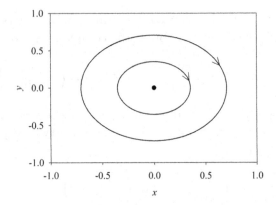

Figure 9.4: The equilibrium (0,0) of system (9.5) is a center with concentric circles as orbits. Centers are neutrally stable.

but this time with increasing amplitude e^t. The equilibrium $(0,0)$ is called an *unstable spiral* (Figure 9.5). The direction of rotation can be determined with a test point as it was for the center. Check that in this example, the spiral is counterclockwise.

9.2.6 Asymptotically Stable Spiral

Consider

$$
\begin{aligned}
x' &= -x + 2y \\
y' &= -5x - 3y.
\end{aligned}
$$

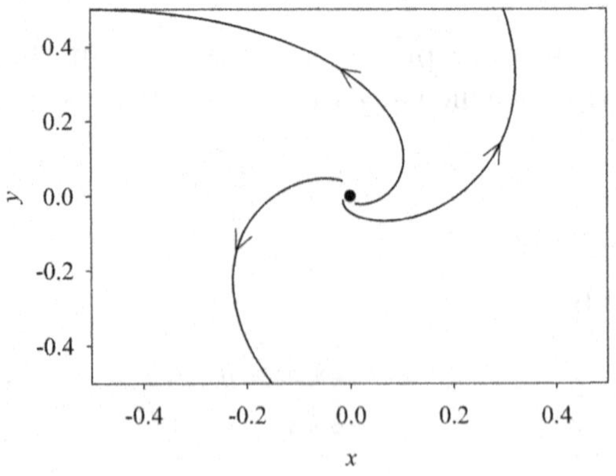

Figure 9.5: The equilibrium (0,0) is an unstable spiral.

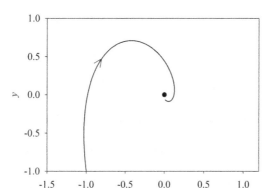

Figure 9.6: The equilibrium (0,0) is an asymptotically stable spiral.

Check that the eigenvalues are $\lambda = -2 \pm 3i$. Note that $e^{(-2\pm3i)t} = e^{-2t}(\cos 3t \pm i \sin 3t)$. In this case, the system oscillates, but with decreasing amplitude. The equilibrium $(0,0)$ is an *asymptotically stable spiral* (Figure 9.6). Check that the direction of rotation is clockwise.

9.2.7 Summary: Eigenvalues Tell All

The six "generic" types of phase plane portraits for two-dimensional linear systems are determined by the eigenvalues as summarized here in Table 9.1 and the corresponding Figure 9.7. For the "nongeneric" other cases, see Henson (2012).

TABLE 9.1 The Six Generic Phase Plane Portraits. (See Figure 9.7)

Figure 9.7	Eigenvalues	Phase Portrait Type
Panel (a)	$\lambda_1 < \lambda_2 < 0$	Asymptotically stable node
Panel (b)	$0 < \lambda_1 < \lambda_2$	Unstable node
Panel (c)	$\lambda_1 < 0 < \lambda_2$	Saddle (unstable)
Panel (d)	$\lambda_{1,2} = \pm bi$	Center (neutrally stable)
Panel (e)	$\lambda_{1,2} = a \pm bi, a < 0$	Asymptotically stable spiral
Panel (f)	$\lambda_{1,2} = a \pm bi, a > 0$	Unstable spiral

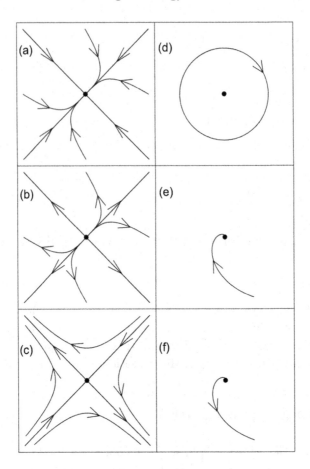

Figure 9.7: The six generic phase plane portraits. (a) Asymptotically stable node. (b) Unstable node. (c) Saddle. (d) Center. (e) Asymptotically stable spiral. (f) Unstable spiral. (See Table 9.1.)

9.3   NONLINEAR SYSTEMS OF ODES

Consider the nonlinear system

$$x' = f(x, y)$$
$$y' = g(x, y)$$

with equilibrium (x_e, y_e). As in Chapter 5, we approximate f and g

with tangent planes near the equilibrium:

$$x' = f(x,y) \approx f(x_e, y_e) + \frac{\partial f}{\partial x}(x_e, y_e)(x - x_e) + \frac{\partial f}{\partial y}(x_e, y_e)(y - y_e)$$

$$y' = g(x,y) \approx g(x_e, y_e) + \frac{\partial g}{\partial x}(x_e, y_e)(x - x_e) + \frac{\partial g}{\partial y}(x_e, y_e)(y - y_e)$$

for $x \approx x_e$ and $y \approx y_e$. Note that $f(x_e, y_e) = 0$ and $g(x_e, y_e) = 0$ because (x_e, y_e) is an equilibrium. Define the variation or displacement from equilibrium by the variables

$$u = x - x_e$$
$$v = y - y_e.$$

Note that $u' = x'$ and $v' = y'$. Thus, near the equilibrium, the variation from equilibrium can be approximated by the linear system

$$u' \approx \frac{\partial f}{\partial x}(x_e, y_e) u + \frac{\partial f}{\partial y}(x_e, y_e) v$$

$$v' \approx \frac{\partial g}{\partial x}(x_e, y_e) u + \frac{\partial g}{\partial y}(x_e, y_e) v,$$

for $u \approx 0$ and $v \approx 0$. This means that near the equilibrium, the system's behavior is determined by the eigenvalues of the coefficient matrix, which is in this case the Jacobian

$$\begin{pmatrix} \frac{\partial f}{\partial x}(x_e, y_e) & \frac{\partial f}{\partial y}(x_e, y_e) \\ \frac{\partial g}{\partial x}(x_e, y_e) & \frac{\partial g}{\partial y}(x_e, y_e) \end{pmatrix}.$$

Let's apply this to an example. Consider the nonlinear system

$$x' = x - xy$$
$$y' = 2y - 2xy.$$

First, we set up the equilibrium equations and factor:

$$x(1 - y) = 0$$
$$2y(1 - x) = 0.$$

From the first equation, $x = 0$ or $y = 1$. If $x = 0$, then from the second equation $y = 0$. If, on the other hand, $y = 1$, then from the second equation $x = 1$. Thus, the two equilibria are $(0, 0)$ and $(1, 1)$. We linearize around each equilibrium in turn. The Jacobian is

$$J = \begin{pmatrix} 1 - y & -x \\ -2y & 2 - 2x \end{pmatrix}.$$

At the equilibrium $(0, 0)$, the Jacobian is

$$J(0, 0) = \begin{pmatrix} 1 & 0 \\ 0 & 2 \end{pmatrix}.$$

The eigenvalues are $\lambda = 1$ with eigenvector $\begin{pmatrix} 1 \\ 0 \end{pmatrix}$ and $\lambda = 2$ with eigenvector $\begin{pmatrix} 0 \\ 1 \end{pmatrix}$. Thus, the $(0, 0)$ equilibrium is an unstable node for the *linear* system, with unstable manifolds on the axes. Therefore, as we shall see, the $(0, 0)$ equilibrium of the *nonlinear* system is also an unstable node, and the phase plane locally around $(0, 0)$ is topologically equivalent to that of the linear system.

At the equilibrium $(1, 1)$, the Jacobian is

$$J(1, 1) = \begin{pmatrix} 0 & -1 \\ -2 & 0 \end{pmatrix}.$$

The eigenvalues are $\lambda = \sqrt{2}$ with eigenvector $\begin{pmatrix} 1 \\ -\sqrt{2} \end{pmatrix}$ and $\lambda = -\sqrt{2}$ with eigenvector $\begin{pmatrix} 1 \\ \sqrt{2} \end{pmatrix}$. Thus, the $(0, 0)$ equilibrium of the *linear* system is a saddle, with unstable and stable manifolds determined by these eigenvectors. Therefore, as we shall see, the $(1, 1)$ equilibrium of the *nonlinear* system is also a saddle, and the phase plane locally around $(1, 1)$ is topologically equivalent to that of the linear system around $(0, 0)$.

Now we can piece the local information together to approximate the phase plane portrait for the nonlinear system (Figure 9.8).

We make these ideas more precise in the next section.

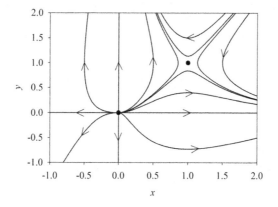

Figure 9.8: Nonlinear phase plane portrait.

9.3.1 Linearization

Definition 9.1 *The **linearization of a nonlinear system***

$$\begin{aligned} x' &= f(x, y) \\ y' &= g(x, y) \end{aligned}$$

at an equilibrium (x_e, y_e) *is the linear system*

$$\begin{aligned} u' &= \frac{\partial f}{\partial x}(x_e, y_e)\, u + \frac{\partial f}{\partial y}(x_e, y_e)\, v \\ v' &= \frac{\partial g}{\partial x}(x_e, y_e)\, u + \frac{\partial g}{\partial y}(x_e, y_e)\, v, \end{aligned}$$

that is,

$$\mathbf{u}' = \mathbf{J}\mathbf{u},$$

where $\mathbf{u} = \begin{pmatrix} u \\ v \end{pmatrix}$ *is the vector of state variables and* $\mathbf{J}$ *is the Jacobian matrix evaluated at the equilibrium* (x_e, y_e).

Definition 9.2 *An equilibrium* (x_e, y_e) *of a system*

$$\begin{aligned} x' &= f(x, y) \\ y' &= g(x, y) \end{aligned}$$

*is **hyperbolic** if and only if all of the eigenvalues of the Jacobian matrix* $J(x_e, y_e)$ *have nonzero real parts.*

Theorem 9.1 *(Linearization Theorem) If $f(x,y)$ and $g(x,y)$ are continuously differentiable in both variables and (x_e, y_e) is a hyperbolic equilibrium of the system*

$$x' = f(x,y)$$
$$y' = g(x,y),$$

then
(1) if all the eigenvalues of the Jacobian $J(x_e, y_e)$ have negative real parts, (x_e, y_e) is asymptotically stable,
(2) if at least one of the eigenvalues of $J(x_e, y_e)$ has positive real part, then (x_e, y_e) is unstable.

The **Hartman-Grobman theorem** says that near hyperbolic equilibria, the nonlinear phase plane portrait is (locally) topologically equivalent to the phase plane portrait of the linearization.

9.4   LIMIT CYCLES, CYCLE CHAINS, AND BIFURCATIONS

If a nonlinear two-dimensional system is *autonomous* (meaning there is no explicit time dependence in the model equations), it can have a limited number of other phase plane configurations that are not possible in autonomous linear systems.

Periodic solutions of two-dimensional systems are associated with closed-loop orbits, called *cycles*, in the phase plane. A famous example that has concentric cycles in the phase plane (that is, a nonlinear center) is the Lotka-Volterra predator-prey model, which we will discuss in the next section. A cycle always surrounds at least one equilibrium. If it attracts nearby orbits, it is called a *limit cycle* (Figure 9.9a). We will see an example of a limit cycle in the Van der Pol oscillator (9.9) below.

Another type of configuration in the phase plane is the *cycle chain* (also known as a *heteroclinic cycle*). This occurs when multiple equilibria are connected by *heteroclinic orbits* (orbits that connect two different equilibria) in a loop structure (Figure 9.9b). Oscillatory solutions may approach a cycle chain, misleading the researcher to think they are approaching a limit cycle. The cycle chain, however, is

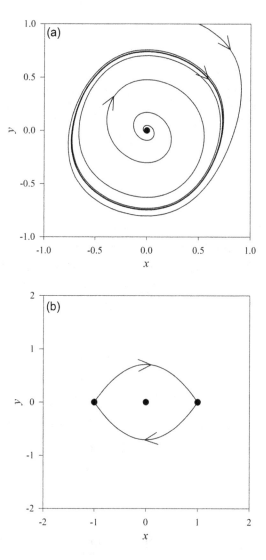

Figure 9.9: (a) Limit cycle of the Van der Pol oscillator (equation 9.9). (b) Cycle chain.

not itself an orbit (although it is an invariant set). You will analyze systems with cycle chains in Exercises 8 and 9.

A *bifurcation* is an abrupt change in phase portrait type that occurs as a parameter is tuned through a critical value. The following example illustrates both limit cycles and bifurcations. Consider the Van der Pol oscillator

$$\begin{aligned} x' &= y - x^3 + ax \\ y' &= -x, \end{aligned} \tag{9.9}$$

in which $a \approx 0$ (a is close to zero). You will show in Exercise 6 that if $a < 0$, the equilibrium is hyperbolic and the nonlinear system has an asymptotically stable clockwise spiral at the origin. If a is "tuned" up to $a = 0$, the real parts of the eigenvalues become zero, so the equilibrium is now non-hyperbolic and the linearization theorem does not apply. Although the linearized system has a center at the origin, you can see from simulations in PPLANE that the nonlinear system still has an asymptotically stable spiral at the origin. If a is further tuned up to $a > 0$, however, the equilibrium again becomes hyperbolic and the nonlinear system has an unstable clockwise spiral at the origin. Thus, the Van der Pol oscillator has a bifurcation at $a = 0$.

For small $a > 0$ in equation (9.9), local trajectories spiral out, away from the origin, yet trajectories farther away from the origin are still spiraling in toward the origin. These trajectories meet in a limit cycle (Figure 9.9a). Thus, as a is tuned from negative to positive values, the limit cycle is born at the origin when $a = 0$ and grows in radius as a increases.

A theorem known as the **Poincare-Bendixson theorem** guarantees that bounded orbits in the phase plane must approach equilibria, cycles, or sets of equilibria connected by heteroclinic and homoclinic orbits.

9.5   LOTKA-VOLTERRA MODELS AND NULLCLINE ANALYSIS

We cannot end this chapter without discussing the famous Lotka-Volterra models of ecology. They usually are studied by means of *nullcline* (or *isocline*) *analysis*. When it is complicated to solve for equilibria and eigenvalues in a nonlinear system, one can determine the direction of the vector field in the phase plane by means of curves of zero growth rate. A nullcline for state variable x is a curve in the phase plane along which $dx/dt = 0$. The intersections of the nullclines for x and y are exactly the equilibria.

9.5.1 Lotka-Volterra Competition

The well-known Lotka-Volterra competition model endowed ecology with the theoretical idea of *competitive exclusion*. If x and y are two competing species, then both have negative interactions with the other. If we assume that each species grows logistically in the absence of the other species and that the two species interact through mass action, the model is

$$\begin{aligned} x' &= x\left(r_1 - a_{11}x - a_{12}y\right) \\ y' &= y\left(r_2 - a_{21}x - a_{22}y\right) \end{aligned} \tag{9.10}$$

with positive coefficients.

We find the nullclines for x by setting

$$0 = x\left(r_1 - a_{11}x - a_{12}y\right),$$

which occurs for $x = 0$ and $y = r_1/a_{12} - a_{11}x/a_{12}$. Along these two lines, the rate of change of x is zero; that is, the vector field is vertical. Similarly, the nullclines for y are the lines $y = 0$ and $y = r_2/a_{22} - a_{21}x/a_{22}$. Along these lines, the rate of change of y is zero, and the vector field is horizontal. These two sets of nullclines have. four possible configurations in phase space (Figure 9.10). In two of the cases, one species always wins (Figure 9.10a and b). In the third case, there is a coexistence equilibrium, but it is a saddle and the winner depends on the initial condition (Figure 9.10c). In the fourth case, the coexistence equilibrium is a stable node and the species coexist (Figure 9.10d).

Coexistence happens when the competition within the species is stronger than the competition between the species (Exercise 10).

9.5.2 Lotka-Volterra Cooperation

In cooperation, both species benefit from the other. If we assume that each species grows logistically in the absence of the other, with a mass action benefit when together, the model is

$$\begin{aligned} x' &= x\left(r_1 - a_{11}x + a_{12}y\right) \\ y' &= y\left(r_2 + a_{21}x - a_{22}y\right) \end{aligned} \tag{9.11}$$

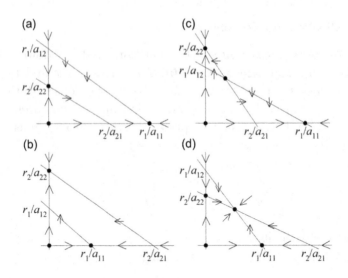

Figure 9.10: Possible phase plane portraits for Lotka-Volterra competition model (9.10). (a) Competitive exclusion: Species x wins. (b) Competitive exclusion: Species y wins. (c) Competitive exclusion: Winner depends on initial condition. (d) Stable coexistence.

with positive coefficients. The nullclines for x are the lines $x = 0$ and $y = -r_1/a_{12} + a_{11}x/a_{12}$, and the nullclines for y are the lines $y = 0$ and $y = r_2/a_{22} + a_{21}x/a_{22}$. Figure 9.11 shows the two possible configurations in the phase plane. In the first case (Figure 9.11a), the cooperators coexist. In the second case, however, the mutual benefit is so strong that it overwhelms the logistic limitations to growth and both species grow without bound (Figure 9.11b)! Robert May famously referred to the second case as an "orgy of mutual benefaction." See Exercise 11.

9.5.3 Lotka-Volterra Predator-Prey

The traditional Lotka-Volterra predator-prey model is

$$x' = x(r_1 - a_{12}y) \tag{9.12}$$
$$y' = y(-r_2 + a_{21}x)$$

with positive coefficients, where x is the prey density and y is the predator density. It is important to note that this model is quite

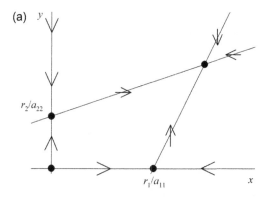

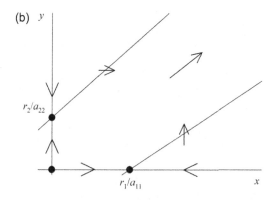

Figure 9.11: Phase plane portraits for Lotka-Volterra cooperation model (9.11). (a) Coexistence; (b) Robert May's "orgy of mutual benefaction."

different from the previous two in that it assumes the prey population grows exponentially in the absence of predators and that the predator population declines exponentially in the absence of prey. Neither of these assumptions is very realistic, because one would expect that (i) both species would have a self-regulatory density-dependent term such as the quadratic nonlinearity in the logistic model, and (ii) the predators would switch to other prey in the absence of species x.

The nullclines for x are the lines $x = 0$ and $y = r_1/a_{12}$, and the nullclines for y are the lines $y = 0$ and $x = r_2/a_{21}$. It is not clear from the nullclines (Figure 9.12) whether the coexistence equilibrium is a spiral or a nonlinear center. It is possible to prove, however, that

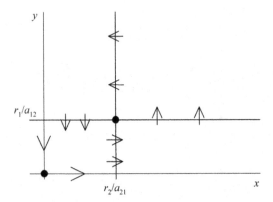

Figure 9.12: Nullclines for the Lotka-Volterra predator-prey system (9.12). It is possible to prove that the orbits are closed curves surrounding the equilibrium, forming a nonlinear center.

the orbits are closed curves and hence surround a nonlinear center. When the periodic solutions are graphed as time series, they exhibit the well-known predator-prey oscillations.

One might consider a more realistic version of the Lotka-Volterra predator-prey model. For example, the Seabird Ecology Team used

$$x' = rx - rx^2/K - axy$$
$$y' = sy - sy^2/C + bxy$$

to model eagle-gull predator-prey dynamics on Protection Island from 1980–2016, where x is the number of gulls and y is the number of eagles (Henson et al. 2019). Here, both populations are assumed to grow logistically in the absence of the other. We showed that the gull population dynamics could be explained by the number of occupied bald eagle territories in Washington with generalized $R^2 = 0.82$. This supported the hypothesis that the rise and decline in gull numbers observed on Protection Island has been due largely to the decline and recovery of the bald eagle population. See Chapter 10, Exercise 9.

9.6 EXERCISES

You will need a computer program to quickly graph phase planes. The free PPLANE Java version (Castellanos and Polking) is very easy to use for this purpose.

1. Obtain equation (9.7) from equation (9.6).

2. Consider the linear system

$$\mathbf{x}' = \mathbf{Ax}. \tag{9.13}$$

For each of the six matrices $\mathbf{A}$ below, do the following calculations by hand:

a. Find the eigenvalues. If the eigenvalues are real:

    i. Find eigenvectors.

    ii. Give two independent eigensolutions of equation (9.13).

    iii. Give the general solution of equation (9.13).

b. Draw the phase plane portrait. If the eigenvalues are real, include the manifolds (orbits of the eigensolutions) in your sketch, along with a few other representative orbits. If the eigenvalues are complex, show the direction of rotation.

c. Classify the $(0,0)$ equilibrium type and give its stability (unstable node, asymptotically stable node, unstable spiral, asymptotically stable spiral, center (neutrally stable), or saddle (unstable)).

d. Check each of your phase portraits using online software such as PPLANE to graph the vector fields.

$$\mathbf{A} = \begin{pmatrix} 3 & -2 \\ 4 & -3 \end{pmatrix}$$

$$\mathbf{A} = \begin{pmatrix} -1 & -2 \\ 2 & -1 \end{pmatrix}$$

$$\mathbf{A} = \begin{pmatrix} -3 & -2 \\ 0 & -2 \end{pmatrix}$$

$$\mathbf{A} = \begin{pmatrix} 0 & 2 \\ -2 & 0 \end{pmatrix}$$

$$\mathbf{A} = \begin{pmatrix} 3 & 2 \\ 1 & 2 \end{pmatrix}$$

$$\mathbf{A} = \begin{pmatrix} 3 & 2 \\ -2 & 3 \end{pmatrix}$$

3. Consider the harmonic oscillator system (9.5).

    a. Prove that equations (9.8) are solutions of system (9.5) for all $c_1, c_2 \in R$.

    b. Prove that solutions (9.8) describe closed curves in the phase plane. In particular, they describe concentric circles of radius $\sqrt{c_1^2 + c_2^2}$ centered on the origin.

4. Consider the nonlinear system

$$\begin{aligned} x' &= 2x + y^2 \\ y' &= x - y. \end{aligned}$$

    a. Find all equilibria.

    b. Find the Jacobian at each equilibrium.

    c. Which of the equilibria are hyperbolic?

    d. For each hyperbolic equilibrium, determine the phase portrait/stability type using the linearization theorem.

    e. Use PPLANE to draw the complete phase portrait for this nonlinear system.

5. Consider the nonlinear system

$$\begin{aligned} x' &= y - x^3 \\ y' &= -x. \end{aligned}$$

    a. Show that $(0,0)$ is the only equilibrium.

    b. Find the Jacobian at $(0,0)$.

    c. Show that $(0,0)$ is **non**hyperbolic and that the linearized equation has a center at $(0,0)$.

    d. Use PPLANE to see that $(0,0)$ is actually an asymptotically stable spiral for the nonlinear equation.

6. Consider the Van der Pol equation from circuits

$$\begin{aligned} x' &= y - x^3 + ax \\ y' &= -x, \end{aligned}$$

where $a \approx 0$.

a. Show that the only equilibrium is $(0,0)$.

b. Find the linearization at $(0,0)$. (Your answer should be a linear system of DEs.)

c. Find the eigenvalues of the Jacobian matrix at $(0,0)$.

d. If $a \gtrsim 0$, what is the phase plane portrait? Note: Let $a \gtrsim 0$ denote positive values of a that are close to zero.

e. If $a \lesssim 0$, what is the phase plane portrait?

f. If $a = 0$, what (if anything) can you conclude from the linearization theorem?

g. Use PPLANE to investigate how the phase plane portrait changes as a ranges from $a = -0.1$ to $a = 0.1$. Use a window size of $-2 < x < 2$ and $-2 < y < 2$. Note the birth of the stable limit cycle (closed-loop solution) as a passes through the critical value $a_{cr} = 0$. This is similar to what's called a *Hopf bifurcation*.

h. For small $a > 0$, can you form a conjecture about the radius of the limit cycle as a function of a?

7. Carry out an isocline analysis of the Van der Pol equation in problem (6). Do a separate analysis for each of $a = 0$, $a \lesssim 0$, and $a \gtrsim 0$.

8. This problem, designed by J. M. Cushing (Cushing 2004), illustrates a cycle chain (heteroclinic cycle). Consider the system

$$x' = (x^2 - 1)y$$
$$y' = (1 - y^2)\left(x + \frac{3}{10}y\right).$$

a. There are seven equilibria. Find them and graph them on the phase plane.

b. Find the Jacobian matrix as a function of x and y.

c. For each of the seven equilibria in turn, find the eigenvalues, linearized phase plane portrait, and stability.

d. Along each of the two vertical lines $x = \pm 1$ in the phase plane, what are the dynamics of y?

e. Along each of the two horizontal lines $y = \pm 1$ in the phase plane, what are the dynamics of x?

f. Graph the nonlinear phase plane portrait. Check your work with PPLANE.

9. This problem illustrates a cycle chain. Part (d) requires integration by separation of variables. If you haven't had second-semester calculus or a course in differential equations, you should review the explanation of this technique in Chapter 7. Consider the system

$$
\begin{aligned}
x' &= y \\
y' &= -x\left(1 - x^2\right).
\end{aligned}
$$

a. There are three equilibria. Find them and place them on the phase plane in a graph.

b. Find the Jacobian matrix as a function of x and y.

c. Show that two of the equilibria are hyperbolic saddles and the third is nonhyperbolic (so the linearization theorem does not apply to it).

d. Prove that all solutions lie on curves of the form $x^2 + y^2 = \frac{1}{2}x^4 + C$. Hint: Compute dy/dx by recalling that $dy/dt = dy/dx \cdot dx/dt$. Use the technique of "separation of variables" to integrate dy/dx.

e. Prove that the nonhyperbolic equilibrium is (locally) a nonlinear center. Hint: Prove that if $C \in (0, 1/2]$, then the equations $x^2 + y^2 = \frac{1}{2}x^4 + C$ describe a family of closed curves containing the origin for $-1 \leq x \leq 1$.

f. Prove that two heteroclinic orbits connect the two saddle equilibria. Hint: Prove that the two saddle equilibria lie on the closed curve associated with $C = 1/2$.

g. Determine whether the trajectories on the closed curves for $C \in (0, 1/2]$ are clockwise or counterclockwise in forward time.

h. Graph the nonlinear phase plane portrait and check your work with PPLANE.

10. For the Lotka-Volterra competition model (9.10), find conditions on the parameters that ensure the possibility of stable coexistence. Give a biological interpretation: Under what conditions is competitive coexistence possible?

11. For the Lotka-Volterra cooperation model (9.11), find conditions on the parameters that lead to unbounded growth (Figure 9.11b). Give a biological interpretation of these conditions.

BIBLIOGRAPHY

Castellanos, J. and Polking, J. C. PPLANE, Java version. https://www.cs.unm.edu/~joel/dfield/. [Freeware for graphing phase plane portraits.]

Cushing, J. M. 2004. *Differential Equations: An Applied Approach.* Prentice-Hall, Upper Saddle River, NJ, p. 308. [A good source for more differential equation exercises and introductory theory, as well as many interesting applied modeling projects using differential equations.]

Henson, S. M. 2012. Phase plane analysis. In A. Hastings and L. Gross (Eds.) *Encyclopedia of Theoretical Ecology* (pp. 538–545). University of California Press, Berkeley. [Includes the "nongeneric" phase plane portraits not covered in this chapter.]

Henson, S. M., Desharnais, R. A., Funasaki, E. T., Galusha, J. G., Watson, J. W., and Hayward, J. L. 2019. Predator-prey dynamics of bald eagles and glaucous-winged gulls at Protection Island, Washington, DC. *Ecology and Evolution* 9:3850–3867. DOI:10.1002/ece3.5011. [Lotka–Volterra-type predator–prey model fitted to gull nest count data and Washington State eagle territory data 1980–2016 with $R^2 = 0.82$, suggesting that gull dynamics were due largely to eagle recovery dynamics. Both theoretical and empirical: models connected to data.]

CHAPTER 10

Seabird Behavior: A Case Study

10.1 WHAT YOU SHOULD KNOW ABOUT THIS CHAPTER

This chapter is based on the initial paper (Henson et al. 2004) of a large body of research carried out by the Seabird Ecology Team, a National Science Foundation-funded research group co-directed by the authors of this textbook. The paper, which was published in *The Auk,* was co-authored with animal behaviorist Joe Galusha and two undergraduate research students. That paper, along with others by the team, is cited in the annotated bibliography for this chapter and can be referred to for more details on this particular project, as well as further information on the team's larger body of work.

The continuous-time modeling techniques in this chapter assume an understanding of the methods of connecting models to data (Chapter 2), including the use of one-step predictions (Chapter 6), as well as familiarity with ordinary differential equations (Chapters 7–9). The chapter also requires the relevant skills in coding (Appendices B–C).

10.2 THE SCIENTIFIC PROBLEM

Wildlife managers and ecologists are interested in temporal changes in numbers of organisms. Numbers of organisms provide baseline information about populations—a starting point for understanding how any ecosystem works.

In the Pacific Northwest region of North America, glaucous-winged gulls (*Larus glaucescens*) are important environmental indicators for the quality of the marine environment, much like the death of canaries indicated the presence of carbon monoxide in coal mines more than a century ago. Gulls are ideal for field study because they are large, abundant, and easily observed. Moreover, they engage in behaviors that are reasonably complex, but not so complex as to make monitoring and analysis difficult.

Recall that Chapter 6 explored the use of mathematical models to understand the dynamics of laboratory populations of flour beetles. A major feature of the beetle work was the rigorous connection of models with actual data, an important step beyond the construction of theoretical models. One of us, Shandelle Henson, was a member of the "Beetle Team." Meanwhile, the other one of us, Jim Hayward, had for several years been collecting time series behavior data in a breeding colony of gulls on Protection Island National Wildlife Refuge, Strait of Juan de Fuca, Washington, the USA (48°07′40″N, 122°55′3″W). When we compared notes, we wondered: If mathematical models could work so well for population dynamics of laboratory insect populations, could they also describe and predict behavioral dynamics of free-living seabirds on Protection Island? We decided to find out.

For a pilot study, we chose a simple system. We decided to see if we could successfully model and predict numbers of glaucous-winged gulls *loafing* (standing, sitting, preening, resting, sleeping) on a pier adjacent to the Protection Island breeding colony. Hayward had a large historical data set spanning four field seasons in which he had monitored hourly numbers of gulls in various habitats, including those loafing on the pier. In the autumn of 2001, we used these data to parameterize and validate an ordinary differential equation model that described fluctuations in gull counts in relation to three environmental variables: day of the year, solar elevation, and tide height. We then used the parameterized model to predict what the counts would be for specific hours on days during the (future) 2002 field season. Along with two undergraduate students, we traveled to Protection Island in the spring of 2002 to test the *a priori* model predictions.

This chapter is the story of how the study was carried out and provides exercises that will allow you to work through the modeling processes we used. Subsequent team papers that use the same methodology are found in the bibliography.

10.3   HISTORICAL DATA

10.3.1   Count Data

The Protection Island gull colony contained more than 3,000 gulls during this study. The gulls move from habitat to habitat during the day. For example, a breeding gull begins the day on its nesting territory. It may fly to the nearby beach and wade into the water for a drink. Then it might fly off to some remote location to feed. When it returns to the island, instead of going directly to its territory, it may land in a communal loafing area to preen and rest. One popular site for this activity is the pier located in a small marina located adjacent to the colony.

One day each week Hayward had made an hourly count of the gulls on the pier during every daylight hour (0600–2100 PDT), May to August, 1997, 1998, 1999, and 2001. Counts were made hourly so that they could be compared to fluctuations in environmental variables such as tide height and solar elevation.

10.3.2   Dividing the Data

The historical count data were assigned randomly to two bins, the first for parameter estimation and the second for independent model validation. In order to preserve variability along the entire tidal cycle, we assigned daily data sets to bins using stratified random sampling: We first divided the 14-day tidal periods into roughly four quarters. We then randomly selected half the daily data sets from each quarter for the parameter estimation bin; the remaining data were placed in the validation bin. The binned historical data are given in Data Set 10.1.

10.3.3   Tide and Solar Elevation Data

Tides in the Strait of Juan de Fuca are "semi-diurnal," meaning there are two high tides and two low tides during each 24-hour period; moreover, one of the low tides is higher than the other low tide. Tidal amplitude changes over a 14-day interval, with the lowest amplitude tide at what we called a "node." (See arrows, Figures 10.2 and 10.4.)

We downloaded historical hourly tide height observations in meters from the National Oceanic and Atmospheric Administration (NOAA) (see bibliography entry for URL) for the Port Townsend, Washington station 9444900, for the days corresponding to the historical pier counts. The tide heights in meters were obtained by multiplying each NOAA observation by 0.93, which is the correction factor for nearby Protection Island. We normalized these corrected hourly tide data by first subtracting the minimal tide value from each tide height in order to set the minimum equal to zero, then dividing each by the resulting maximum to set the new maximum equal to one, and finally by adding one so that the result was between one and two:

$$T(t) = \frac{\text{tide} - \min{(\text{tide})}}{\max{(\text{tide} - \min{(\text{tide})})}} + 1. \tag{10.1}$$

This resulted in a non-dimensional tide height function with $1 \leq T(t) \leq 2$.

Similarly, we obtained hourly solar elevations from NOAA (see bibliographic entry for URL). We set negative solar elevations to zero and then performed a normalization equivalent to equation (10.1) in order to produce a non-dimensional solar elevation function with $1 \leq S(t) \leq 2$.

10.4   GENERAL MODEL

We used an ordinary differential equation compartmental model to track the dynamics on the pier. The net rate of change in the number of gulls N loafing on the pier is the rate at which birds land on the pier (inflow rate) minus the rate at which birds leave the pier (outflow rate):

$$\frac{dN}{dt} = (\text{Inflow rate}) - (\text{Outflow rate}).$$

We assumed that:

(H1) Fluctuations in numbers of gulls on the pier occurred in response to an environmental variable $E(t)$.

(H2) Numbers of gulls on the pier during daylight hours could be described using a two-compartment model: one compartment consisting of the loafing area and the other compartment consisting of all other locations.

(H3) Gulls landed on the pier at a per capita rate proportional to $E(t)$, and they left the pier at a per capita rate inversely proportional to $E(t)$.

(H4) The total number of gulls in the two compartments depended on the time of year and was proportional to the weekly maximal number of gulls landing on the pier.

We let $N(t)$ be the number of birds on the pier at time t, where t is the day of the year plus the decimal fraction of the time within that day. We took $\beta K_p(t)$ to be the total number of gulls in the system, both on the pier and everywhere else, at time t, where $K_p(t)$ estimates the maximum number of gulls on the pier anytime during the year. With these definitions, the expression $\beta K_p(t) - N(t)$ represents the number of all gulls in the system that are not on the pier at time t. The inflow rate is the per capita inflow rate multiplied by the number of birds in the system at other locations, that is, $\alpha E(t) (\beta K_p(t) - N)$. The outflow rate is the per capita outflow rate multiplied by the number of birds on the pier, that is, $N/(\alpha E(t))$. The constants of proportionality $\alpha, \beta > 0$ are the parameters that we needed to estimate from the historical data. The maximum number of gulls in the adjacent colony was approximately 60 times the maximum value of $K_p(t)$; consequently, we selected 80 as a generous upper limit for β. The model is therefore

$$\frac{dN}{dt} = \alpha E(t) (\beta K_p(t) - N) - \frac{N}{\alpha E(t)}, \tag{10.2}$$

where $0 < \alpha < \infty$ and $1 \le \beta \le 80$.

We estimated $K_p(t)$ from pier count data taken throughout January 1 to March 21 and May 23 to December 31 during 1997, 1998, 1999, and 2001. Using these occupancy data, we estimated the seasonal maximum number of gulls for the fluctuating pier counts by fitting a modified lognormal curve to the means of the maximal pier counts for every week of the year:

$$K_p(t) = 76.36 \exp \left[\frac{[\ln (40.29 - t/7) - 2.504]^2}{-0.7225} \right].$$

After day 275 and before day 65 of the following year, the mean maximum pier counts were zeros or ones; consequently, we set $K_p(t) = 0$ for those intervals. See Figure 10.1.

10.5 ALTERNATIVE MODELS

We tested three alternative hypotheses for (H1) on the historical data:

(H1a) $E(t) = T(t)$

(H1b) $E(t) = 1/S(t)$

(H1c) $E(t) = T(t)/S(t)$

The three alternative hypotheses substituted into equation (10.2) produce three alternative models: the "tidal model," the "solar model," and the "tidal-solar model."

10.6 MODEL PARAMETERIZATION

Using the method of conditioned least squares (CLS), we estimated parameters for each of the three alternative models. For each hour, a one-step prediction of the next observation was conditioned on the previous observation. Specifically, for a set of $n + 1$ successive hourly

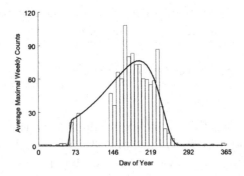

Figure 10.1: Maximal historical counts. The height of each bar is the mean historical maximal count for the pier for that week, averaged over the years 1997–1999 and 2001. The curve is the fitted function $K_p(t)$. (Originally published in *The Auk* 121:382. Used with permission of Oxford University Press.)

observations $\{x_0, x_1, \ldots, x_n\}$, the model was integrated numerically to produce n hourly one-step predictions $\{y_1, y_2, \ldots, y_n\}$, where y_{i+1} denotes the model prediction at time $i + 1$, given the observation x_i at time i as the initial condition for the model. The differences $x_i - y_i$ between the observed and predicted values at time i yielded the conditioned one-step residual errors. The residual sum of squares

$$\text{RSS}(\theta) = \sum_{i=1}^{n} (x_i - y_i)^2$$

was minimized as a function of the vector θ of model parameters. The vector of CLS parameter estimates for the model was the minimizer $\hat{\theta}$. We produced the one-step predictions using the MATLAB *ode45* integrator, and we minimized $\text{RSS}(\theta)$ with the Nelder-Mead algorithm *fminsearch* in MATLAB.

We used the generalized R^2 value given by

$$R^2 = 1 - \frac{\text{RSS}(\hat{\theta})}{\sum_{i=1}^{n} (x_i - \bar{x})^2}$$

as a measure of goodness of fit, where $\bar{x}$ denotes the sample mean of the observations $\{x_1, x_2, \ldots, x_n\}$. The value R^2 estimates the proportion of the observed variability explained by the model, thereby providing a measure of model prediction accuracy.

Computational note: In order to numerically integrate equation (10.2), which depends on the environmental functions T and/or S, one must spline hourly tides and solar elevations in order to produce smooth curves that give between-hour values for T and S. See Exercise 3.

CLS parameter estimates and R^2 values for the three alternative models as fitted to the estimation data are shown in Table 10.1. You will reproduce these values in Exercise 7.

10.7 MODEL SELECTION

In this study, it was not necessary to use the AIC to select the best model, because all three alternative models had the same number

TABLE 10.1 CLS Parameter Estimates and R^2 Values
for the Three Alternative Models as Fitted to the
Estimation Data

Model	$\hat{\alpha}$	$\hat{\beta}$	$\hat{\sigma}^2$	R^2
H1a	0.31	2.0	244	0.38
H1b	0.10	80	225	0.43
H1c	0.35	2.8	166	0.58

Note that for (H1b), the local minimizer for RSS occurs
on the boundary of allowed parameter space with $\beta = 80$.

of parameters. In this situation, the model with the lowest AIC
corresponds to the model with the highest R^2 (Exercise 4).

Based on the R^2 values in Table 10.1, we eliminated the tidal
and solar models in favor of the tidal-solar model. Tidal-solar model
simulations are displayed with the estimation data in Figure 10.2.

10.8 MODEL VALIDATION

Using the reserved historical validation data in Data Set 10.1, we
independently evaluated the performance of the best model without
re-estimating its parameters (Exercise 8). The results are in Table 10.2.

The $R^2 = 0.60$ for the validation data is comparable to $R^2 = 0.58$
for the estimation data. This similarity supported confidence in the
model validation outcome. Figure 10.2 displays the model predictions
for the validation data (without refitting).

10.9 TEST OF *A PRIORI* PREDICTIONS

The analysis described above occurred during the autumn of 2001. We
decided to use our parameterized and validated model to see if we
could predict what would happen in the gull colony at a future date.

TABLE 10.2 Model (H1c) Validation
Results

	Estimation Data	Validation Data
R^2	0.58	0.60

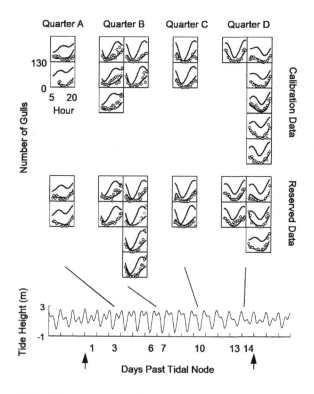

Figure 10.2: Model prediction (lower curve), hourly historical observations (circles), and tide height (upper curve) for the estimation
data and the reserved validation data. Each panel represents one day,
with a horizontal scale of 5–20 hours PST, which is 6–21 hours PDT.
A typical tidal curve for Protection Island is shown at the bottom.
Tidal nodes are indicated with arrows. Data from days occurring
during the same quarter of the tidal sequence are stacked vertically.
(Originally published in *The Auk* 121:383. Used with permission of
Oxford University Press.)

We chose the upcoming breeding season: May and June of 2002. In
particular, we used the tidal-solar model with parameters in Table 10.1
to generate hourly predictions for numbers of gulls loafing on the pier
during daylight hours for May and early June 2002 (Figures 10.3 and
10.4). The model predictions showed periodicities at three temporal
scales: high-frequency daily oscillations (Figure 10.4c), the expected

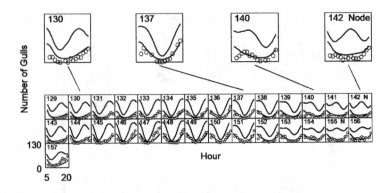

Figure 10.3: Model (H1c) prediction (lower curve), hourly data from 2002 (circles), and tide height (upper curve). Each daily panel is identified with the day of the year and has a horizontal scale of 5–20 hours PST, which is 6–21 hours PDT. Tide height is graphed on a vertical scale of −1 to 3 m. (Originally published in *The Auk* 121:385. Used with permission of Oxford University Press.)

low-frequency seasonal fluctuation (Figure 10.4a), and an unexpected medium-frequency oscillation (Figure 10.4c) that coincided with a biweekly tidal pattern (Figure 10.4e, arrows). The form and minimal values of the predicted daily fluctuations depended in an unexpected way on the time within the tidal cycle relative to the solar cycle (Figures 10.3 and 10.4c). On some days, for example, model predictions oscillated out of phase with the tidal cycle (Figure 10.3, days 142 and 155), an unexpected result in light of previous studies in the literature.

We began collecting data to test the model predictions at 0900 on May 9, 2002. Thereafter, data collection continued at hourly intervals from 0600 to 2100 PDT (dawn to dusk) until 2100 on June 6, 2002. Our counts of gulls on the pier and associated structures were made from a 30-meter-high bluff located approximately 100 m from the pier. If there was a human disturbance on the pier during the count or within 30 minutes before the count, that count was eliminated.

The 2002 data, given in Data Set 10.2, contained all three predicted periodicities (Figure 10.4b and d). Model predictions closely approximated the count data (Figure 10.3) both qualitatively and quantitatively, with a goodness of fit of $R^2 = 0.66$. During the

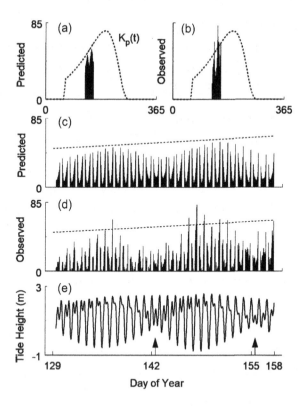

Figure 10.4: Model (H1c) predicted oscillations on three different time scales. (a and c) Model predictions for the spring of 2002, shown with the seasonal pier envelope (dotted curve). Oscillations are present on daily, biweekly, and yearly scales. (b and d) Data observations corresponding to the predictions in (a and c). (e) Tidal oscillation for the data collection time period in 2002. The tidal nodes are indicated with arrows. (Originally published in *The Auk* 121:386. Used with permission of Oxford University Press.)

first several days of each internodal period, daily minimal counts lagged morning low-low tides (e.g., Figure 10.3, day 130), in mid-period shifted to a point below midday low-low tides (e.g., Figure 10.3 day 137), later in the period preceded afternoon low-low tides (e.g., Figure 10.3, day 140), and at the node coincided with high tide (e.g., Figure 10.3, day 142). After the node, daily maximal counts increased and then decreased toward the next node in a reflection of the biweekly oscillation predicted by the model (Figure 10.4b and d). Finally, from the first to the second biweekly oscillation period, the average daily counts increased, which reflected the predicted increase in counts during the spring (Figure 10.4b and d).

10.10   STEADY-STATE MODEL

The prediction of a differential equation model typically is determined by integration over the past. But the gull-pier system recovers quickly (empirically, in less than one hour) following a disturbance, and it can be demonstrated that the state of the system from hour to hour depends more on the current environmental conditions at time t than on the past history of the system (Henson, Hayward, and Damania 2006). Said another way, the parameterized tidal-solar model exhibits rapid transient dynamics (rapid in comparison with its steady-state dynamics).

For such systems with fast transient recovery times, it is possible to use a technique called *multiple time scale analysis*, which produces two algebraic models: one for the transient dynamic and one for the steady-state dynamic. A classic reference for multiple time scale analysis is the book by Lin and Segel (1988). Here is an idea of how it works. We begin with the differential equation

$$\frac{dN}{dt} = \alpha E(t) \left(\beta K_p(t) - N(t) \right) - \frac{N(t)}{\alpha E(t)},$$

and we consider the fast time scale $\tau = t/\varepsilon$, where $\varepsilon > 0$ is a small number. Then using chain rule, we obtain

$$\frac{dN}{dt} \frac{dt}{d\tau} = \frac{dN}{d\tau} = \alpha E(\tau) \left(\beta K_p(\tau) - N(\tau) \right) - \frac{N(\tau)}{\alpha E(\tau)},$$

and so

$$\varepsilon \frac{dN}{dt} = \alpha E(\tau) \left(\beta K_p(\tau) - N(\tau) \right) - \frac{N(\tau)}{\alpha E(\tau)}.$$

For small enough ε, we have

$$0 \approx \alpha E(\tau) \left(\beta K_p(\tau) - N(\tau) \right) - \frac{N(\tau)}{\alpha E(\tau)}. \tag{10.3}$$

Using $E(\tau) = T(\tau)/S(\tau)$, solving for $N(\tau)$, and renaming τ as t, we obtain the fast time scale (steady state) solution $N_0(t)$, which is the algebraic model (Exercise 5)

$$N_0(t) = \frac{\beta K_p(t)}{1 + \frac{1}{\alpha^2}\left(\frac{S(t)}{T(t)}\right)^2}. \tag{10.4}$$

The ability to make model predictions of the steady-state dynamics of a system using this approach can be useful in some applications, including resource management. Although the algebraic equation (10.4) is simpler to use than the differential equation, *it is unable to account for transient dynamics following disturbances* and cannot make *one-step* predictions for model fitting. In Exercise 6, you will compute the R^2 for the steady-state algebraic model (10.4) on the 2002 data (Data Set 10.2), using the tidal-solar model parameters from Table 10.1.

10.11 DISCUSSION

10.11.1 Importance of Scale

So, is animal behavior largely deterministic or largely stochastic? The answer is: It depends on the scale.

Fluctuation in animal numbers over the short term indicates a variety of competing functional needs. For example, during a single day an individual bird may loaf, feed, and defend its territory against intruders. To carry out all these behaviors, the bird must move from habitat to habitat. Physiological limitations and environmental variables are important in opening and closing opportunities for these behaviors. It is not too surprising, then, that at the group level, some behaviors are fairly deterministic, as we have seen in this chapter.

Nevertheless, animals such as gulls display high levels of individual variation in behavior. An individual gull decides to move from one habitat to another within a unique set of historical contingencies. While it is true that a decision to change habitats may be influenced by what other gulls do, the decision may also made without this influence. We saw no evidence of coaction or social facilitation in arrivals and departures from the pier, except in cases when the pier was disturbed by human activity. Instead, gulls landed on the pier and left the pier individually throughout the day, suggesting that the decision by individual gulls to loaf may be more or less independent of the behavior of the other gulls. The results in this chapter therefore suggest that deterministic forces at some group scales can be more important than stochastic individual variability in a system.

The 1989 Robert H. MacArthur Award Lecture, presented by Simon Levin in Toronto, Canada, and appearing in print in the journal *Ecology* as the paper "The Problem of Pattern and Scale in Ecology," is a classic discussion of scale (Levin 1992). It should be read by everyone who is interested in the dynamics of ecological systems.

10.11.2 Resource Management

Dynamic models of ecological systems are needed for proper understanding and management of natural resources. The successful prediction of the dynamics of organisms in their natural environments can lead to an understanding of food production, conservation, the spread of disease, and resource management. Mathematical models can generate testable hypotheses about natural systems. But this can happen only if data are rigorously connected to models through parsimonious modeling assumptions and model parameterization using appropriate statistical methodologies, model selection from a set of alternative models, and model validation. Data sampling must be sufficiently dense to allow for model parameterization and validation. Field sampling must occur at intervals smaller than relevant environmental periodicities. It is important that models, once parameterized, be validated on independent data sets to avoid simple curve fitting, which has little explanatory power. Finally, the most useful mathematical models are those that make successful *a priori* predictions of future system dynamics, especially if the predictions are unexpected.

The Seabird Ecology Team has successfully applied the methodology described in this chapter to a number of other systems. See the annotated bibliography for this chapter.

10.12 EXERCISES

1. Obtain from NOAA the Port Townsend hourly tidal observations in meters for hours 0600–2100 PDT on July 2, 1997. Correct these values for Protection Island with a factor of 0.93 as explained in this chapter. Do your values match those in Data Set 10.1?

2. Using all of the tidal observations and solar elevations in Data Set 10.1, compute hourly normalized values of $T(t)$ and $S(t)$ with equation (10.1). Do your values match those in Data Set 10.1?

3. Consider the vector of hours $\mathbf{t} = (0, 1, 2, 3, 4, 5)$ and a corresponding vector of function values $T(\mathbf{t}) = (1.0, 0.54, -0.42, -0.99, -0.65, 0.28)$. Use a cubic spline interpolator to estimate $T(2.1)$ and $T(3.8)$. Hint: If you are using MATLAB, you can use the cubic spline interpolator $spline(t, T, \tau)$ in which t is the vector of times at which you already know the function value, T is the vector of function values at those times, and τ is the time (or vector of times) at which you want to estimate the interpolated value of the function.

4. Prove: If each model in a set of alternative one-dimensional models has the same number of parameters, and if the residuals are independent and Gaussian with mean zero and constant variance σ^2, then choosing the model with the smallest AIC is equivalent to choosing the one with the largest R^2. Hint: See Exercise 18 in Chapter 2.

5. In equation (10.3), replace $\approx$ with $=$ and solve for N to obtain equation (10.4).

6. How well does the steady-state algebraic model (10.4) with $\alpha = 0.35$ and $\beta = 2.8$ (H1c; Table 10.1) fit the data in Data Set 10.2? Compute the R^2. Show the fit visually with appropriate graphs of daily data together with predictions. Attach your programs, input files, and output.

7. Write the programs to estimate parameters for model (10.2) on the estimation data. Reproduce the parameter estimates and R^2 values in Table 10.1. Attach your programs, input files, and output. Hint: Use $(0.1, 10)^\top$ as the initial guess for the parameter vector. Print the parameters to the screen as they iterate. Answers may differ slightly depending on the software you use.

8. Reproduce the R^2 values in Table 10.2. Attach your programs, input files, and output.

9. Two-dimensional ODE parameterization project: This problem concerns an eagle-gull predator-prey study (Henson et al. 2019). In that paper, the Seabird Ecology Team fitted a Lotka-Volterra-type predator-prey model to gull nest count data and Washington State eagle territory data 1980–2016 with $R^2 = 0.82$, suggesting that gull dynamics were due largely to eagle recovery dynamics. The paper is open access and contains the observed data.

   a. Read the paper (Henson et al. 2019).

   b. Reproduce the equilibrium and stability analysis given in Appendix A of Henson et al. (2019).

   c. Following the methods in that paper, reproduce the parameter estimates in Table 2 of the paper and the R^2 values.

BIBLIOGRAPHY

Cowles, J. D., Henson, S. M., Hayward, J. L., and Chacko, M. W. 2013. A method for predicting harbor seal (*Phoca vitulina*) haulout and monitoring long-term population trends without telemetry. *Natural Resource Modeling* 26:605–627. DOI: 10.1111/nrm.12015. [Applies the differential equation approach and time scale analysis methods of this chapter to harbor seal haul-out behavior. Both theoretical and empirical: models connected to data.]

Damania, S. P., Phillips, K. W., Henson, S. M., and Hayward, J. L. 2005. Habitat patch occupancy dynamics of glaucous-winged gulls (*Larus glaucescens*) II: A continuous-time model. *Natural Resource Modeling* 18:469–499. [Applies the differential equation approach and time scale analysis methods of this chapter to the movement of gulls within a system of three habitat patches dedicated to loafing. Both theoretical and empirical: models connected to data.]

Hayward, J. L., Henson, S. M., Logan, C. J., Parris, C. R., Meyer, M. W, and Dennis, B. 2005. Predicting numbers of hauled-out harbour seals: a mathematical model. *Journal of Applied Ecology* 42:108–117. [Applies the differential equation approach and time scale analysis methods of this chapter to harbor seal haul-out behavior. Both theoretical and empirical: models connected to data.]

Hayward, J. L., Henson, S. M., Tkachuck, R., Tkachuck, C., Payne, B. G., and Boothby, C. K. 2009. Predicting gull/human conflicts with mathematical models: a tool for management. *Natural Resource Modeling* 22:544–563. [Empirical test of this chapter's loafing model for glaucous-winged gulls loafing at different locations during various stages of the breeding season in different years, and for herring (*Larus argentatus*) and great black-backed gulls (*L. marinus*) loafing on roof tops on Appledore Island, Maine, the USA. Both theoretical and empirical: models connected to data.]

Henson, S. M., Dennis, B., Hayward, J. L., Cushing, J. M., and Galusha, J. G. 2007. Predicting the dynamics of animal behaviour in field populations. *Animal Behaviour* 74:103–110. DOI: 10.1016/j.anbehav.2006.11.015. [Presents a general mathematical framework for modeling animal behaviors and habitat patch occupancies with compartmental differential equations. Both theoretical and empirical: models connected to data.]

Henson, S. M., Desharnais, R. A., Funasaki, E. T., Galusha, J. G., Watson, J. W., and Hayward, J. L. 2019. Predator-prey dynamics of bald eagles and glaucous-winged gulls at Protection Island, Washington, USA. *Ecology and Evolution* 9:3850–3867. DOI:10.1002/ece3.5011. [Lotka-Volterra-type predator-prey model fitted to gull nest count data and Washington State eagle territory data 1980–2016 with $R^2 = 0.82$, suggesting that gull dynamics were due largely to eagle recovery dynamics. Both theoretical and empirical: models connected to data.]

Henson, S. M., Galusha, J. G., Hayward, J. L., and Cushing, J. M. 2007. Modeling territory attendance and preening behavior in a seabird colony as functions of environmental conditions. *Journal of Biological Dynamics* 1:95–107. DOI: 10.1080/17513750601032679 [Applies compartmental modeling methods of this chapter to behaviors of territory attendance and preening in a gull colony, where each behavior is considered a compartment and the flow rates between compartments are functions of environmental variables. Both theoretical and empirical: models connected to data.]

Henson, S. M. and Hayward, J. L. 2010. The mathematics of animal behavior: an interdisciplinary dialogue. *Notices of the American Mathematical Society* 57:1248–1258. [Perspective piece on the integration of biology and mathematics and the Seabird Team's approach to modeling animal behavior.]

Henson, S. M., Hayward, J. L., Burden, C. M., Logan, C. J., and Galusha, J. G. 2004. Predicting dynamics of aggregate loafing behavior in gulls at a Washington colony. *Auk* 121:380–390. DOI: 10.1642/0004-8038(2004)121[0380:PDOALB]2.0.CO;2. [The paper on which this chapter is based. Both theoretical and empirical: models connected to data.]

Henson, S. M., Hayward, J. L., and Damania, S. P. 2006. Identifying environmental determinants of diurnal distribution in marine birds and mammals. *Bulletin of Mathematical Biology* 68:467–482. DOI: 10.1007/s11538-005-9009-0. [Describes the time scale technique mentioned in this chapter. Both theoretical and empirical: models connected to data.]

Henson, S. M., Weldon, L. M., Hayward, J. L., Greene, D. J., Megna, L. C., and Serem, M. C. 2012. Coping behaviour as an adaptation to stress: Post-disturbance preening in colonial seabirds. *Journal of Biological Dynamics* 6:17–37. DOI: 10.1080/17513758.2011.605913. [Differential equation approach incorporating logistic regression and Darwinian dynamics to investigate theoretically how a behavior with crucial physiological function might evolve into a comfort behavior. Both theoretical and empirical: models connected to data.]

Levin, S. A. 1992. The problem of pattern and scale in ecology. *Ecology* 73:1943–1967. DOI: 10.2307/1941447. [Classic paper on scale in ecology, presented as the Robert H. MacArthur Award Lecture in 1989 in Toronto, Canada. Every student of ecology and mathematical biology should read this paper.]

Lin, C. C. and Segel, L. A. 1988. *Mathematics Applied to Deterministic Problems in the Natural Sciences.* SIAM, Philadelphia. [Classic text in applied mathematics and mathematical biology. Belongs on every mathematical biologist's bookshelf.]

Moore, A. L., Damania, S. P., Henson, S. M., and Hayward, J. L. 2008. Modeling the daily activities of breeding colonial seabirds: Dynamic occupancy patterns in multiple habitat patches. *Mathematical Biosciences and Engineering* 5:831–842. DOI: 10.3934/mbe.2008.5.831. [Applies methods of this chapter to multiple habitat patches. Both theoretical and empirical: models connected to data.]

National Oceanic and Atmospheric Administration. Solar Position Calendar. https://gml.noaa.gov/grad/solcalc/azel.html.

National Oceanic and Atmospheric Administration. Tides and Currents. https://tidesandcurrents.noaa.gov/waterlevels.html?id=9444900.

Payne, B. G., Henson, S. M., Hayward, J. L., Megna, L. C., and Velastegui Chavez, S. R. 2015. Environmental constraints on haul-out and foraging dynamics in Galpágos marine iguanas. *Journal of Coupled Systems and Multiscale Dynamics* 3:208–218. DOI: 10.1166/jcsmd.2015.1077. [Applies the methods in this chapter to haulout in marine iguanas. Both theoretical and empirical: models connected to data.]

IV

Regression Models

CHAPTER 11

Introduction to Regression

11.1 WHAT YOU SHOULD KNOW ABOUT THIS CHAPTER

This chapter briefly introduces basic ideas from regression theory. Regression models are statistical models, but they are also mathematical models that we parameterize with data, hence their place in this book. We omit many useful tools in regression theory, but we hope this brief, basic treatment will give the student a conceptual foundation for further investigation.

11.2 LINEAR REGRESSION

We begin with linear regression, which students may encounter in high school mathematics courses. The idea is to put a "best-fit line" through data points of the form (x, y) that have an approximately linear trend when y is graphed against x. This is called "regressing y against x." For example, we might regress crop yield against rainfall. Here, crop yield is the outcome, which is the dependent variable. The dependent variable is also called the *response* variable in regression. Rainfall is the *factor*, which is the independent variable. The independent variable is also called a *predictor* variable or *explanatory* variable in regression. Note the order of words in the terminology "regress y against x," and "the response of y to x." If we are thinking in terms of graphs, this is equivalent to the way mathematicians say "graph y against x." Always

DOI: 10.1201/9781003265382-15

say "vertical against horizontal." Many people mistakenly reverse the order.

11.2.1 Simple Linear Regression (Single Factor)

Suppose we wish to know how a dependent variable y depends on a factor x, under the assumption that x and y have a linear relationship. The mathematical statement of this assumption is

$$y = \beta_0 + \beta_1 x,$$

where the *coefficients* β_0 and β_1 are the model parameters. Here, β_1 is the *slope* and β_0 is the *intercept*.

The intercept β_0 quantifies the value of y when $x = 0$. Let us consider the interpretation of the slope β_1. Suppose factor x changes by c units. Then the resulting change in y is

$$\begin{aligned} \Delta y &= y_{\text{new}} - y_{\text{old}} \\ &= (\beta_0 + \beta_1(x+c)) - (\beta_0 + \beta_1 x) \\ &= \beta_1 c. \end{aligned}$$

That is, if x changes by c units, then y changes by $\beta_1 c$ units. In particular, if x changes by one unit, then y changes by β_1 units. Thus, β_1 is the rate of change of y with respect to the factor x. This is, of course, consistent with our understanding from calculus that slope is equivalent to the derivative, which is the rate of change. Note that if β_1 is positive, then y increases as x increases, whereas if β_1 is negative, then y decreases as x increases.

11.2.2 Multiple Linear Regression (Multiple Factors)

Suppose now that y depends linearly on n factors $x_1, x_2, \ldots, x_n$. The *multiple regression* (linear regression with multiple factors) model is

$$\begin{aligned} y &= \beta_0 + \beta_1 x_1 + \beta_2 x_2 + \ldots + \beta_n x_n \qquad (11.1) \\ &= \beta_0 + \sum_{j=1}^{n} \beta_j x_j. \end{aligned}$$

If factor x_i changes by c units, *with all other factors held constant*, then it is straightforward to check that the resulting change in y is (Exercise 1)

$$\Delta y = \beta_i c.$$

Thus, coefficient β_i is the rate of change of y with respect to factor x_i *when all other factors are held constant.* The sign of the coefficient β_i determines whether y is positively or negatively correlated with factor x_i.

11.2.3   Stochastic Model and Parameter Estimation

A stochastic version of model (11.1) in which Gaussian noise is additive is

$$Y = \beta_0 + \sum_{j=1}^{n} \beta_j x_j + \sigma\varepsilon,$$

where $\sigma > 0$ is a parameter representing the standard deviation of the noise and ε is a standard normal random variable (mean zero and standard deviation one). The residuals are

$$res = y - \left(\beta_0 + \sum_{j=1}^{n} \beta_j x_j \right),$$

where y is an observed value (realization of the random variable Y) associated with factors having values $(x_1, x_2, \ldots, x_n)$. The residuals are realizations of the random variable $\sigma\varepsilon$. That is, this model assumes that the residuals are normally distributed about zero with standard deviation σ.

The likelihood that an observed data point is a realization of the stochastic model, that is, the likelihood that the residual comes from the hypothesized distribution of noise, is

$$\frac{1}{\sigma\sqrt{2\pi}} e^{-\frac{1}{2}\left(\frac{res}{\sigma}\right)^2}.$$

Assuming the observations are independent, the likelihood L that all of the residuals generated from the data set come from the hypothesized distribution is

$$L = \prod_{\text{data}} \frac{1}{\sigma\sqrt{2\pi}} e^{-\frac{1}{2}\left(\frac{res}{\sigma}\right)^2}.$$

The ML parameter estimates are those that maximize the likelihood L.

Parameter estimation routines for regression equations are available in the libraries of scientific computing programs. In MATLAB, for example, one can use the function *glmfit*.

11.2.4 Confidence Intervals for Regression Coefficients

Suppose a regression slope coefficient is estimated to be $\widehat{\beta}$, with standard error SE. The standard error of a parameter estimate is the standard deviation of its sampling distribution. That is, if you resampled the data many times and considered the distribution of the estimated values $\widehat{\beta}$, the standard deviation of that distribution would be SE. The endpoints of the 95% confidence interval for β are

$$\widehat{\beta} \pm 1.96 \times \text{SE}.$$

11.3 LOGISTIC REGRESSION

Logistic regression (Hosmer and Lemeshow 2000) is a technique commonly used to quantify the effect of factors on a *binary* variable (an event occurs, or it does not occur). Let P be the probability of the event. Then P takes on values between zero and one, making linear regression inappropriate for predicting P because regression lines predict values from negative infinity to positive infinity. Thus, the dependent variable P must be transformed so that its range is R.

The *odds* of the event are defined to be

$$\frac{P}{1 - P}.$$

Note that, whereas P takes on values between zero and one, the associated odds take on values between zero and infinity. The *log odds* of the occurrence of the event are given by

$$\ln\left(\frac{P}{1 - P}\right),$$

which takes on values between negative infinity and positive infinity.

In logistic regression, the log odds are regressed on a vector $x = (x_1, x_2, \ldots, x_n)$ of factors:

$$\ln\left(\frac{P}{1 - P}\right) = \beta_0 + \sum_{j=1}^{n} \beta_j x_j, \tag{11.2}$$

where P is the probability of the event. The intercept β_0 calibrates the baseline log odds of the event when all factors are zero, and the regression coefficients $\beta = (\beta_1, \ldots, \beta_n)$ quantify the response of the log odds to changes in the factors.

11.3.1 Odds Ratios (ORs)

The coefficients β have a convenient interpretation. If factor x_i increases by c units while all other factors remain constant, then the log odds changes by $\beta_i c$ units:

$$\ln\left(\frac{P_2}{1-P_2}\right) - \ln\left(\frac{P_1}{1-P_1}\right)$$

$$= \left(\beta_0 + \beta_i(x_i + c) + \sum_{\substack{j=1 \\ j \neq i}}^{n} \beta_j x_j\right) - \left(\beta_0 + \sum_{j=1}^{n} \beta_j x_j\right)$$

$$= \beta_i c,$$

where P_1 and P_2 are the probabilities of the event before and after the change, respectively. Thus, by the laws of logarithms, the *odds ratio*, denoted OR, is

$$\text{OR} = \frac{P_2/(1-P_2)}{P_1/(1-P_1)} = e^{\beta_i c}.$$

This means that, given an increase in factor x_i by $c > 0$ units, *with all other factors held constant,* the odds of the event are $e^{\beta_i c}$ times what they were before.

If $\beta_i > 0$, then the odds ratio is greater than one (OR > 1), meaning the odds of the event have increased. If $\beta_i < 0$, the odds ratio is less than one (OR < 1), and the odds of the event have decreased. For example, if OR $= 1.25$, the interpretation is that the odds of the event increase 25% with a c-unit increase in x_i. If OR $= 0.85$, the odds decrease 15% with a c-unit increase in x_i. The researcher chooses a convenient value of the reference c so that the results are intuitively clear. Note that these statements of association do not necessarily imply causation.

11.3.2 OR Confidence Intervals

Odds ratios typically are presented with confidence intervals. The 95% confidence interval for $OR = e^{\beta c}$ is

$$\left(e^{c(\beta - 1.96 \times SE)}, e^{c(\beta + 1.96 \times \mathrm{SE})} \right),$$

where SE is the standard error of the regression coefficient β. If the confidence interval includes the value of 1, this indicates that we are not confident whether $OR < 1$ or $OR > 1$, and hence, we cannot claim that the odds of the event significantly decrease or increase with an increase in the associated factor.

Many other useful topics, such as how to rank factors according to their importance, are found in the definitive books by Burnham and Anderson (Burnham and Anderson 2002) and by Hosmer and Lemeshow (Hosmer and Lemeshow 2000).

11.4 GENERALIZED LINEAR MODELS (GLMs)

As we saw above, logistic regression is a form of linear regression in which a log-odds transformation of the dependent variable is regressed against factors. In general, one can regress a transformed dependent variable $g(y)$ against the factors $x = (x_1, x_2, \ldots, x_n)$:

$$g(y) = \beta_0 + \sum_{j=1}^{n} \beta_j x_j. \tag{11.3}$$

The transformation g is called the *link function*, and the model is called a *generalized linear model* (GLM). Common examples of GLMs include logistic regression for binary data (0 or 1 outcomes), Poisson regression for count data, gamma regression for exponential response data, and linear regression for linear response data. The link functions for these types of regression are shown in Table 11.1.

For more information on GLMs, see the definitive book by McCullagh and Nelder (1989).

11.5 INTERACTION TERMS

The right-hand side of equation (11.3) can also include *interaction terms* of the form $\beta_{ij} x_i x_j$. These terms are necessary if the effect of

TABLE 11.1   Link Functions for Regression

Regression Type	Link Function	Type of Data (Dependent Variable)
Gamma	$g(y) = -y^{-1}$	Exponential response data, $(0, \infty)$
Linear	$g(y) = y$	Linear response data $(-\infty, \infty)$
Logistic	$g(y) = \ln(y/(1-y))$	Binary data (yes/no), $\{0, 1\}$
Poisson	$g(y) = \ln y$	Count data, $\{0, 1, 2, \ldots\}$

one factor depends on the value of another factor. In such a case, the meaning of the coefficients β_i and the formulas for the confidence intervals change. For example, consider the logistic regression model

$$\ln(\text{odds}) = \beta_0 + \beta_1 x_1 + \beta_2 x_2 + \beta_{12} x_1 x_2.$$

If factor x_1 is increased by c units with all other factors held constant, then the change in log odds is

$$\begin{aligned}
\Delta \ln(\text{odds}) &= (\beta_0 + \beta_1(x_1 + c) + \beta_2 x_2 + \beta_{12}(x_1 + c)x_2) \\
&\quad - (\beta_0 + \beta_1 x_1 + \beta_2 x_2 + \beta_{12} x_1 x_2) \\
&= \beta_1 c + \beta_{12} c x_2.
\end{aligned}$$

Thus, the odds ratio

$$\text{OR} = e^{c(\beta_1 + \beta_{12} x_2)}$$

depends on the value of factor x_2.

11.6   EXERCISES

1. Prove that if factor x_i increases by c units in equation (11.1), and all other factors are held constant, then the resulting change in y is $\beta_i c$.

2. Consider the Poisson regression model

$$\ln y = \beta_0 + \beta_1 x_1 + \beta_2 x_2 + \beta_3 x_3.$$

If factor x_1 increases by c units, and all other factors are held constant, then how does y change?

3. Consider the GLM

$$g(y) = \beta_0 + \beta_1 x_1 + \beta_2 x_2,$$

where g is the link function for gamma, logistic, or Poisson regression (Table 11.1). For each of these three types of models, solve the regression equation for y.

4. Consider the following multiple regression model that predicts VO2max from AGE in years, WEIGHT in kilograms, and heart rate (HRT) in beats/minute in men (adapted from https://statistics.laerd.com/stata-tutorials/multiple-regression-using-stata.php):

$$\text{VO2max} = 101.04 - 0.17\text{AGE} - 0.39\text{WEIGHT} - 0.12\text{HRT}.$$

   a. What predicted change in VO2max results from adding 2 extra kg of weight, with all other factors held constant?

   b. What is the predicted change in VO2max if the heart rate goes up by 3 beats/min, with all other factors held constant?

5. Consider the logistic regression model

$$\ln(odds) = \beta_0 + \beta_1 x_1 + \beta_2 x_2 + \beta_3 x_3.$$

   a. If the slope coefficient for factor x_3 is $\beta_3 = -0.1$ and you increase x_3 by $c = 5$ units, with all other factors held constant, what is the odds ratio (OR)?

   b. If x_3 increases by five units, with all other factors held constant, then the odds of the outcome decrease by what percent?

6. Second-hand smoke increases the risk of lung cancer in nonsmokers and is a serious health risk. Suppose the odds ratio for lung cancer for a nonsmoker living with a smoker versus not living with a smoker is about OR = 1.28, all other factors held constant. Living with a smoker (vs. not living with a smoker) increases a nonsmoker's odds of developing lung cancer by what percent?

7. Consider the outcome of passing a class (0 or 1) as a function of the average number of study hours per week (STUDY) and the average number of hours of sleep per night (SLEEP).

a. Use the following fictional data set to parameterize the logistic regression model

$$\ln\left(\text{odds of passing}\right) = \beta_0 + \beta_1 x_1 + \beta_2 x_2.$$

Use the MATLAB library function *glmfit(X, Y, 'binomial')* or another GLM fitting function in the programming language you are using.

PASS	STUDY (x_1)	SLEEP (x_2)
0	6	4
1	5	8
1	3	4
0	1	5
1	6	5
0	0	4
1	6	6
1	4	7
1	5	8
0	1	7
0	1	9
1	7	7

b. Fill in the missing values in the results table using the given reference value c.

Factor	b_i	c	SE	OR	95% CI
STUDY		0.5 hours			
SLEEP	0.3391	2 hours	0.4960	1.970	$(0.2819, 13.77)$

c. Interpret the results in words, in terms of the reference value c, the odds ratio OR, and significance. For example, the interpretation for SLEEP would be: "For each extra 2 hours of sleep per night, the odds of passing the class increased by 97%. However, this trend was not significant given that the OR confidence interval included numbers both above and below one, due to the large standard error."

BIBLIOGRAPHY

Burnham, K. P. and Anderson, D. R. 2002. *Model Selection and Multi-Model Inference: A Practical Information-Theoretic Approach*, 2nd ed. Springer-Verlag, New York. [Comprehensive and user friendly text on methods of model selection.]

Henson, S. M., Weldon, L. M., Hayward, J. L., Greene, D. J., Megna, L. C., and Serem M. C. 2012. Coping behaviour as an adaptation to stress: Post-disturbance preening in colonial seabirds. *Journal of Biological Dynamics* 6:17–37. DOI: 10.1080/17513758.2011.605913. [Incorporates regression models into differential equation models for animal behavior.]

Hosmer, D. W. and Lemeshow, S. 2000. *Applied Logistic Regression*, 2nd ed. John Wiley & Sons, New York. [The classic logistic regression text.]

McCullagh, P. and Nelder, J. A. 1989. *Generalized Linear Models*, 2nd ed. Chapman & Hall/CRC Press, Boca Raton, FL. [Definitive text on GLMs.]

CHAPTER 12

Climate Change and Seabird Cannibalism: A Case Study

12.1   WHAT YOU SHOULD KNOW ABOUT THIS CHAPTER

This chapter is based on a large study conducted by the Seabird Ecology Team. The team collaborated with mathematician Lynelle Weldon to examine the relationship between a climate-driven environmental variable (sea surface temperature) and egg cannibalism in gulls. Results from this study were first published in *The Condor: Ornithological Applications* (Hayward et al. 2014). The team produced several additional related publications, which are listed in the annotated bibliography for this chapter.

The methodology in this chapter employs basic logistic regression (Chapter 11) and coding skills (Appendix B).

The Condor study (Hayward et al. 2014) was comprehensive and used a number of techniques we will not explain here, except perhaps in a cursory fashion, such as Akaike weights, model-averaged parameter estimation (Burnham and Anderson 2002), design variables, measures of overdispersion (Hosmer and Lemeshow 2000), and advances in techniques for validating logistic regression models (Giancristofaro and Salmaso 2003). In the exercises, you will use basic logistic regression without these supporting techniques, and the parameters you obtain will therefore differ somewhat from those obtained in *The Condor* paper (Hayward et al. 2014). That paper, however, might serve as

DOI: 10.1201/9781003265382-16

a useful resource for approaching a comprehensive logistic regression analysis in the reader's own research projects.

12.2   THE SCIENTIFIC PROBLEM

Over the course of many field seasons at a colony of several thousand glaucous-winged gulls (*Larus glaucescens*) at Protection Island National Wildlife Refuge, Washington, the USA, we noticed large accumulations of broken eggshell littering a few of the nesting territories. We watched the owners of these territories invade the territories of fellow residents, grab an egg, fly the egg back to their own territories, and eat the contents. Eggshell fragments from the stolen eggs accumulated on the territories of these egg cannibals (Figure 12.1).

A cannibal is any animal that, like these gulls, kills and eats members of its own species, and their victims can be at any stage of the life cycle, including the egg stage. Members of at least 1,300 species are known to engage in cannibalism (Polis 1981), although the real number is certainly much higher. Cannibals are found among zooplankton, arthropods, fish, reptiles, birds, mammals, and other animal groups. Once thought to be an aberration, cannibalism is now considered a normal life history trait of many species (Elgar and Crespi 1992; Fox 1975).

Cannibalism exerts important influences on ecological and evolutionary features of animals. It can, for example, influence energy relations among members of ecological communities, reduce their reproductive success, change the sizes of their populations, modify their social behaviors, promote kin selection, and even result in complex

Figure 12.1: (a) Male cannibal holding egg in bill. (b) Accumulation of eggshell fragments on the territory of an egg cannibal. (Photos by James Hayward.)

nonlinear population dynamics such as chaos (Cushing, Henson and Hayward 2015; Elgar and Crespi 1992; Polis 1981).

Crowding, odd behavior patterns by subdominant individuals, psychological and physiological stress, and the availability of victims can elicit cannibalism (Fox 1975). The size, age, sex, habitat, and developmental stage of potential victims also may play roles. Moreover, some individuals are more genetically predisposed than others to engage in cannibalistic behavior. Most commonly, however, lack of food or poor food quality leads to cannibalism (Dong and Polis 1992).

Ornithologists have long known that gulls cannibalize their neighbors' eggs, a behavior reported for at least 19 gull species (Polski et al. 2021). The behavior, however, has not been well studied, and most such reports have come from incidental observations during research into other aspects of gulls' lives. We wondered about the extent and impact of this behavior on a breeding colony, so we decided to examine this phenomenon at Protection Island. One of the papers that resulted from this project provides the first comprehensive assessment of the behavioral ecology of cannibalism by gulls (Polski et al. 2021).

Only males on the Protection Island colony steal eggs. A cannibalistic male locates an egg to steal in one of two ways. He may fly slowly over an undisturbed colony looking for a momentarily unprotected nest, or he may wait for an eagle to fly over, disturbing residents in large areas of the colony so that they fly up from their nests and leave their eggs exposed. In either case, the cannibal capitalizes on the opportunity, quickly lands by the exposed nest, and grabs an egg. The cannibal then flies the stolen egg back to his territory, gently holding the egg in his bill or swallowing it into his crop whole. Once back on his territory, the cannibal drops the egg or regurgitates it from his crop, pecks it open, and devours the slimy contents. If his mate is present, she typically begs to be allowed to participate in the feeding. Most of the time, the male shares his loot with his mate, but occasionally, he pecks at her and refuses to share.

Eating two eggs per day just about fulfills the energy requirements of an adult gull (Hayward et al. 2014), but some cannibals in our study stole more than two eggs per day. In June 2014, for example, one male stole 81 eggs in 30 days, and another took 75 eggs during the same period (Polski et al. 2021). Unlike most colony residents, super egg predators like these spent little time looking for food off the colony.

We compared the contents of regurgitated food pellets of non-cannibals and egg cannibals. (Gulls regurgitate pellets of undigestible material just like owls.) Not surprisingly, pellets from cannibals were significantly more likely to contain fragmented eggshell than pellets from non-cannibals, but pellets from cannibals were also significantly less likely to contain fish remains than those of non-cannibals. This suggested that cannibals were substituting eggs for fish as their primary protein source.

Only about 1 in a 100 of the territories on the Protection Island gull colony were defended by egg cannibals, yet these few cannibals stole 1 out of every 9 eggs produced in the entire colony in both 2014 and 2015. Despite the fact they are few in number, egg cannibals impact the colony in disproportionate ways, and several birds at Protection Island exhibited this behavior over multiple years.

How did the reproductive output of egg cannibals compare with those of fellow colony residents? Egg cannibals produced significantly fewer eggs than non-cannibals. Moreover, the proportion of eggs taken from the nests of cannibals by *other* cannibals was significantly higher than the proportion of eggs taken from the nests of non-cannibals. So, egg cannibalism appears to be a feeding tactic used by less attentive and less successful breeders.

Marine birds associated with the Pacific Ocean experience food shortages every few years as a result of El Nino-Southern Oscillation (ENSO) events. These events raise the sea surface temperature (SST), lower the thermocline (the depth at which the temperature of water suddenly becomes cooler), and weaken upwellings (which bring nutrients up to higher levels in the water column). As a result of these changes, plankton become less abundant toward the sea surface, and this drives forage fish to lower levels in the water column. Gulls, too buoyant to dive, have fewer fish to eat during these times and must travel longer distances to find enough food for themselves and their young. All this leads to lower reproductive success, which may lead to population declines.

We wondered whether ENSO-related rises in SST impacted the rate of egg cannibalism in a breeding colony of glaucous-winged gulls.

12.3 DATA

We collected data from late May to mid-July, 2006–2011, at the glaucous-winged gull colony at Protection Island National Wildlife Refuge, Washington, the USA. Protection Island (48°07′40″N, 122°55′3″W) is located in the southeast corner of the Strait of Juan de Fuca. The gull colony contained more than 2,400 breeding pairs during this study.

We selected five rectangular sample plots (Figure 12.2, Plots A–E), with a combined area of 4,205 m^2, and with a combined total of 199–267 sample nests per year. Each plot was checked every day in the late afternoon. When the first egg was laid in a nest, a numbered stake was placed near the nest and the egg was labeled as the A-egg. Thereafter, the nest was monitored each afternoon. Each subsequent egg was labeled and recorded as well. The ultimate fate of each egg was determined and recorded as cannibalized, eagle predated, addled (dead during incubation), died during hatching, hatched, or other (punctured, nest flooded, or rolled out of nest). We distinguished cannibalized eggs from eagle-predated eggs by the fact that cannibals steal eggs and fly

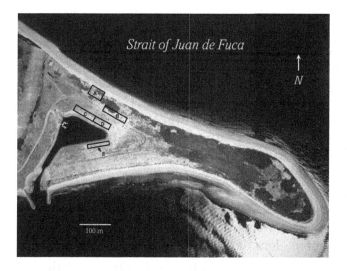

Figure 12.2: Violet point on Protection Island, showing five sample areas A–E in the gull colony.

them back to their own territories, whereas eagles eat eggs where they find them, leaving distinctive patterns of broken eggshell behind.

We measured the distance from the center of each sample nest to the center of the nest of its nearest neighbor. We recorded the nest as residing in one of four habitats: (i) short or sparse vegetation (SV), (ii) tall dune grass (TG), (iii) beside a shrub or log (SL), or (iv) beach (BC). In 2007–2011, we also determined the mass of each egg on the day it was laid.

The average SST from September to May prior to each breeding season was determined using data from the Port Townsend, Washington buoy (PTWW1), a floating National Oceanic and Atmospheric Administration (NOAA) structure located 12 km east of Protection Island. The interval from September to May was chosen for two reasons: (i) Breeding success in gulls is determined in part on resource availability before egg production, and (ii) the effects of changes in the physical environment on seabird populations are not immediate (Smith et al. 2017). Warmer-than-average SSTs preceded the 2007 and 2010 breeding seasons due to El Nino conditions.

The data for 2006–2011 are given in Data Set 12.1.

12.4 LOGISTIC REGRESSION ANALYSIS

The fates of 2,932 eggs monitored over the five breeding seasons from 2007 to 2011 were assessed by logistic regression in terms of whether eggs were cannibalized or not (1 or 0). Cannibalism was considered as a function of SST, egg mass (MASS), nearest neighbor distance (NN), number of days before or after the mean laying date for the season (DAYS), sample area (PLOT; five values, A–E), habitat type (HAB; four values, SV, SL, TG, and BC), egg order as laid in the nest (ORDER), and the total number of eggs laid in the clutch (CSIZE). The log odds of cannibalism were regressed on these eight factors and on 17 interaction terms between PLOT and HAB (three of the 20 possible combinations did not occur) with an intercept term.

A suite of alternative models was obtained from the global model by taking all submodels, that is, all possible linear combinations of the eight factors, with intercepts. For models with both PLOT and HAB variables, the interaction terms were included. We determined parameter estimates and the Akaike information criterion (AIC) for the global model and all submodels.

The best model, that is, the one with the smallest AIC, included all of the factors except ORDER, NN, and MASS. Because the best model did not include MASS, which was the only measurement not taken in 2006, in the exercises you will analyze the entire 6-year (2006–2011) data set (Data Set 12.1).

12.5 MODEL VALIDATION

We then validated the selected model. Because regression models measure trends but are not mechanistically based, the values of R^2 (both fitted and validation) are typically quite low. Thus, for regression modeling the goodness-of-fit measure R^2 used throughout the rest of this text is often uninformative. Indeed, model validation techniques for logistic regression models are an ongoing subject of research.

In this study (Hayward et al. 2014), we used a validation technique by Giancristofaro and Salmaso (Giancristofaro and Salmaso 2003), in which the egg data were split into two random samples: one for parameter estimation (75% of the data) and one for validation (25% of the data). The selected model was fitted to the estimation data, and the model performance was then measured on both samples. This process was repeated 100 times. Details are found in Giancristofaro and Salmaso (2003) and Hayward et al. (2014).

12.6 OUTCOMES

The results of the analysis described above are summarized in Table 12.1, which shows only those variables found to be significant. The parameters used to compute the ORs in Table 12.1 were not the fitted parameters for the best model, but rather were the "model-averaged" parameter estimates. A model-averaged parameter estimate is the average over all models (in the suite of alternative models) that contain that parameter, weighted by its Akaike weight. We do not discuss Akaike weights or model-averaged parameter estimates here; details can be found in Burnham and Anderson (2002).

The most important thing to understand in this section is how to interpret the results in Table 12.1. Make sure you can derive the following interpretations from Table 12.1.

TABLE 12.1 Cannibalism Odds Ratios (ORs) with 95% Confidence Intervals (CI) Associated with a c Unit Increase in the Factor or Relative to the Given Reference Variable

Factor	c or Reference Value	OR	95% CI
SST	0.1 deg	1.10	(1.06, 1.13)
DAYS	1 day	1.09	(1.06, 1.11)
CSIZE	1 egg		
2 eggs		0.13	(0.07, 0.21)
>2 eggs		0.09	(0.05, 0.15)
PLOT B	HAB SV		
HAB SL		0.28	(0.23, 0.33)
HAB TG		0.06	(0.04, 0.09)
PLOT C	HAB SV		
HAB SL		0.54	(0.45, 0.63)
HAB BC		0.68	(0.62, 0.75)
PLOT D	HAB SV		
HAB SL		0.28	(0.26, 0.32)
HAB BC		0.40	(0.34, 0.48)
PLOT E	HAB SV		
HAB SL		0.39	(0.35, 0.42)
HAB BC		0.68	(0.54, 0.87)

Habitats are short or sparse vegetation (SV), beside shrub or log (SL), beach (BC), and beside or in tall grass (TG). Only significant variables and interactions are listed here. For the complete results, see Hayward et al. (2014).

1. The odds of cannibalism increased 10% for each 0.1°C rise in SST, with all other factors held constant.

2. The odds of cannibalism for an egg increased by 9% for each day it was laid away from the mean laying date, with all other factors held constant.

3. An egg in a three-egg nest was 91% less likely to be cannibalized than an egg from a one-egg nest, and an egg from a two-egg nest was 87% less likely to be cannibalized than an egg from a one-egg nest (all other factors held constant).

4. Compared to an egg in a nest in short or sparse vegetation (SV), an egg in a nest beside a shrub or log (SL) was the least likely to be cannibalized, except in Plot B, in which eggs in tall grass

(TG) were even less likely to be cannibalized. In Plots C, D, and E, eggs in nests on the beach (BC) also were less likely to be cannibalized than those in short or sparse vegetation (SV), but they were more likely to be cannibalized than those beside a shrub or log (SL), all other factors held constant.

12.7 CLIMATE CHANGE, CANNIBALISM, AND REPRODUCTIVE SYNCHRONY

The fitness of individuals and the dynamics of populations are directly affected by cannibalism (Dong and Polis 1992). Cannibalism of juveniles by adults may allow population survival when resource levels are low, followed by a redirection of reproductive effort to times when resources are more available (Elgar and Crespi 1992; Henson 1997).

Cannibalism is especially common among adult gulls when food supplies are low, although ours was the first study to directly link egg cannibalism with environmental conditions associated with low food supply.

During our research, egg cannibalism became more common when the SST increased. In particular, a 0.1°C rise in SST was associated with a 10% increase in the odds that an egg was cannibalized. When SST is high, as during El Nino events, plankton and fish, primary food items for gulls, drop to lower, cooler levels in the water column. Unlike many seabirds, gulls cannot dive, so food is more difficult to obtain during these times; eggs, nutritious and conveniently abundant, become a more frequent food source for colony residents.

The highest degrees of egg cannibalism during our study occurred during the 2007 and 2010 breeding seasons, both of which were preceded by higher than average SSTs in months prior to the breeding seasons. Egg cannibalism during these times provided relatively low-cost, locally available food for hungry breeding birds. Indeed, as already noted, some gulls cannibalize more than two eggs per day during the incubation period, probably most of the caloric intake by these birds during this time (Polski et al. 2021).

Sea surface temperatures in the Strait of Juan de Fuca surrounding Protection Island increased by approximately 1°C from 1950 to 1998, and this warming trend likely will continue in the face of global warming. Our data suggest this warming trend may increase the rate of egg cannibalism.

It seems, however, that gulls have evolved a way to reduce the impact of cannibalism on their reproductive efforts. During the years of low SST and low egg cannibalism, egg-laying occurs during a relatively short period, which overwhelms would-be egg and chick predators from outside the colony with a glut of food; the chance that a given egg is predated by bald eagles is reduced (the "Fraser Darling effect" (Darling 1938)). By contrast, during the years of high SST and high egg cannibalism, the egg-laying period lengthens and females nesting close together tend to synchronize their every-other-day egg laying until they achieve the typical three-egg clutch; by laying their eggs on the same day as their neighbors, the chance that a given egg is predated by a cannibal neighbor is reduced (Henson et al. 2010). In other words, gulls switch between two breeding tactics, depending on whether the threats to their offspring are from outside (eagles) or inside (cannibals) the colony (Weir et al. 2020).

Under the lead of one of our colleagues, Gordon Atkins, we found that females tended to avoid copulation on egg-laying days, but welcomed copulation on the intervening days. The very loud, pulsating copulation call of the males functioned as the synchronizing signal (Atkins, Hayward, and Henson, 2021; Atkins et al. 2017).

Our studies of cannibalism raise many unanswered questions. For example, are some gulls genetically predisposed to cannibalize their neighbors' eggs, or is this primarily a learned behavior? Are eggs with certain colors and pigmentation patterns more vulnerable to cannibalism than others? Do egg cannibals exhibit other forms of antisocial behavior outside the breeding season? Will increasing SSTs associated with climate change increase the incidence of egg cannibalism on gull colonies? Hopefully, these and other questions will be answered by future research.

Of the Seabird Team papers in the annotated bibliography below, some are theorem-proof mathematics papers and some are empirical papers. In terms of topics, they address various aspects of cannibalism and egg-laying synchrony in the context of climate change. The annotations will help the reader know how each paper relates to this chapter.

12.8 EXERCISES

1. Fit model

$$\log(\text{odds}) = b_0 + b_1\text{SST} + b_2\text{DAYS} + b_3\text{CSIZE}$$

(no interaction terms) to Data Set 12.1 using logistic regression, where *odds* is the odds an egg is cannibalized. In this exercise, let CSIZE be the actual number of eggs in the clutch, which is slightly different from the definition of CSIZE in Hayward et al. (2014). Also compute odds ratios (ORs) and 95% confidence intervals for the ORs. Use the *glmfit(x,y,'binomial')* function in MATLAB or other software for logistic regression.

a. Fill in the missing entries in this results table, given the reference *c*.

Factor	b	c	SE	OR	95% CI
SST		0.1 deg			
DAYS		1 day			
CSIZE	−0.5671	1 egg	0.0686	0.5672	$(0.4958, 0.6489)$

b. Interpret the results in words, in terms of the reference value *c*, the odds ratio OR, and significance. For example, the interpretation for CSIZE could be: "For each extra 1 egg in the clutch, the odds that a given egg is cannibalized decreases by 43%. This decreasing relationship is significant because the entire 95% confidence interval for OR is less than one."

c. Give a short biological discussion of the results. For example, for CSIZE you might say: "Eggs from a larger clutch are less likely to be cannibalized. There are at least two possible reasons for this: (i) with more eggs in a clutch, each egg is less likely to be taken during a predation event, and (ii) parental investment in a full clutch is higher, leading to more careful guarding of the nest by parents."

d. Compare your results to those in Table 12.1, which are taken from Hayward et al. (2014) and are based on model-averaged parameter estimates.

2. Fit model

$$\log(\text{odds}) = b_0 + b_1 \text{SST}$$

to Data Set 12.1 using logistic regression, where *odds* is the odds an egg is cannibalized. Use the *glmfit(x,y, 'binomial')* function in MATLAB or other software for logistic regression.

a. Fill in the following table:

Factor	b	c	SE	OR	95% CI
SST		0.1 deg			

b. Interpret the results in words, in terms of the reference value c, the odds ratio OR, and significance.

c. Compare your results to those in Table 12.1, which are taken from Hayward et al. (2014).

3. Fit model

$$\log(\text{odds}) = b_0 + b_1 \text{DAYS}$$

to Data Set 12.1 using logistic regression, where *odds* is the odds an egg is cannibalized. Use the *glmfit(x,y, 'binomial')* function in MATLAB or other software for logistic regression.

a. Fill in the following table:

Factor	b	c	SE	OR	95% CI
DAYS		1 day			

b. Interpret the results in words, in terms of the reference value c, the odds ratio OR, and significance.

c. Compare your results to those in Table 12.1, which are taken from Hayward et al. (2014).

4. Complete Exercises 1–3. Suppose, in general, that there are two important factors X_1 and X_2. Discuss in depth the difference between regressing on each factor individually and regressing on both factors at once.

BIBLIOGRAPHY

Atkins, G. J., Hayward, J. L., and Henson, S. M. 2021. How do gulls synchronize every-other-day egg laying? *Wilson Journal of Ornithology* 133:226–235. DOI: 10.1676/20-00019. [The connection between socially-facilitated mounting and egg-laying synchronization. Both theoretical and empirical.]

Atkins, G. J., Reichert, A. A., Henson, S. M., and Hayward, J. L. 2017. Copulation call coordinates timing of head-tossing and mounting behaviors in neighboring glaucous-winged gulls (*Larus glaucescens*). *Wilson Journal of Ornithology* 129:560–567. DOI: 10.1676/16-004.1. [Mounting and copulation calls spread through the colony via social facilitation. Empirical.]

Atkins, G. J., Sandler, A. G., McLarty, M., Henson, S. M., and Hayward, J. L. 2015. Oviposition behavior in glaucous-winged gulls (*Larus glaucescens*). *Wilson Journal of Ornithology* 127:486–493. DOI: 10.1676/14-151.1. [First detailed description of egg-laying behavior in gulls. Empirical.]

Burnham, K. P. and Anderson, D. R. 2002. *Model Selection and Multi-Model Inference: A Practical Information-Theoretic Approach*, 2nd ed. Springer-Verlag, New York. [Comprehensive and user friendly text on information-theoretic methods of model selection.]

Burton, D. and Henson, S. M. 2014. A note on the onset of synchrony in avian ovulation cycles. *Journal of Difference Equations and Applications* 20:664–668. DOI: 10.1080/10236198.2013.870564. [Mathematical treatment of egg-laying synchrony as an adaptive response to egg cannibalism in seabirds. Based on the master's thesis of Danielle Burton. Theoretical.]

Cushing, J. M. and Henson, S. M. 2018. Periodic matrix models for seasonal dynamics of structured populations with application to a seabird population. *Journal of Mathematical Biology* 77:1689–1720. DOI: 10.1007/s00285-018-1211-4. [Mathematical analysis of a general class of discrete-time matrix models that account for changes in behavioral tactics within the breeding season and their dynamic consequences at the population level across breeding seasons. Application to cannibalism and reproductive synchrony in gulls. Theoretical.]

Cushing, J. M., Henson, S. M., and Hayward, J. L. 2015. An evolutionary game theoretic model of cannibalism. *Natural Resource Modeling* 28:497–521. DOI: 10.1111/nrm.12079. [Uses matrix models and bifurcation theory to investigate population and evolutionary dynamic consequences of adult-on-juvenile cannibalism. In the presence of cannibalism, a population can survive under circumstances of low resource availability which, in the absence of cannibalism, lead to extinction. The evolutionary version of the model shows that cannibalism can be an evolutionarily stable strategy (ESS). Presents a gull egg cannibalism model as an application. Theoretical.]

Darling, F. 1938. *Bird Flocks and the Breeding Cycle.* Cambridge University Press, Cambridge. [Presents the "Fraser Darling Effect" hypothesis as to why large colonies of birds breed simultaneously during a short breeding season.]

Dong, Q. and Polis, G. A. 1992. The dynamics of cannibalistic populations: A foraging perspective. In M. A. Elgar & B. J. Crespi (Eds.) *Cannibalism: Ecology and Evolution Among Diverse Taxa* (pp. 13–37). Oxford University Press, Oxford. [Classic reference on cannibalism.]

Elgar, M. A. and Crespi, B. J. 1992. Ecology and evolution of cannibalism. *Journal of Evolutionary Biology* 7:1–12. [Classic reference on cannibalism.]

Fox, L. R. 1975. Cannibalism in natural populations. *Annual Review of Ecology and Systematics.* 6:87–106. DOI: 10.1146/annurev.es. 06.110175.000511. [Classic reference on cannibalism.]

Gallos, D., Gallos, C., Watson, W., and Henson, S. M. 2018. A note on synchronous egg laying in a seabird behavior model. *Journal of Difference Equations and Applications* 24:1953–1966. DOI: 10.1080/10236198.2018.1544633. [Mathematical treatment of egg-laying synchrony as an adaptive response to egg cannibalism in seabirds. Based on the undergraduate Honors theses of Christiana Gallos and Dorothea Gallos. Theoretical.]

Giancristofaro, R. A. and Salmaso, L. 2003. Model performance analysis and model validation in logistic regression. *Statistica* 63:375–396. DOI: 10.6092/issn.1973-2201/358. [Presents a validation method for logistic regression models.]

Hayward, J. L., Weldon, L. M., Henson, S. M., Megna, L. C., Payne, B. G., and Moncrieff, A. E. 2014. Egg cannibalism in a gull colony increases with sea surface temperature. *The Condor: Ornithological Applications* 116:62–73. DOI: 10.1650/CONDOR-13-016-R1.1. [Paper on which this chapter is based, showing that a 0.1 degree rise in sea surface temperature is associated with a 10% increase in the odds that an egg is cannibalized. Both theoretical and empirical: models connected to data.]

Henson, S. M. 1997. Cannibalism can be beneficial even when its mean yield is less than one. *Theoretical Population Biology* 51:109–117.

DOI: 10.1006/tpbi.1997.1303. [Mathematical treatment of benefits of cannibalism. Theoretical.]

Henson, S. M., Cushing, J. M., and Hayward, J. L. 2011. Socially-induced ovulation synchrony and its effect on seabird population dynamics. *Journal of Biological Dynamics* 5:495–516. DOI: 10.1080/17513758.2010.529168. [Poses a discrete-time model of socially-stimulated ovulation synchrony. Shows mathematically that synchrony can increase total population size and allow the population to persist at lower birth rates. Both theoretical and empirical: models connected to data.]

Henson, S. M., Hayward, J. L., Cushing, J. M., and Galusha, J. G. 2010. Socially induced synchronization of every-other-day egg laying in a seabird colony. *Auk* 127:571–580. DOI: 10.1525/auk.2010.09202. [Announcement of first demonstration of egg-laying synchrony. Both theoretical and empirical: models connected to data.]

Hosmer, D. W. and Lemeshow, S. 2000. *Applied Logistic Regression*, 2nd ed. John Wiley & Sons, New York. [The classic logistic regression text.]

McWilliams, K. M., Sandler, A. G., Atkins, G. J., Henson, S. M., and Hayward, J. L. 2018. Courtship and copulation in glaucous-winged gulls, *Larus glaucescens*, and the influence of environmental variables. *Wilson Journal of Ornithology* 130:270–285. DOI: 10.1676/16-151.1. [Detailed description of courtship and copulation behavior based on the master's thesis of Kelly McWilliams. Empirical.]

Nurhan, Y. I. and Henson, S. M. 2021. Cannibalism and synchrony in seabird egg-laying behavior. *Natural Resource Modeling* 34:e12325. DOI: 10.1111/nrm.12325. [Mathematical treatment of egg-laying synchrony as an adaptive response to egg cannibalism in seabirds. Based on the undergraduate Honors thesis of Yosia Nurhan. Theoretical.]

Polis G. A. 1981. The evolution and dynamics of intraspecific predation. *Annual Review of Ecology and Systematics.* 12:225–251. DOI: 10.1146/annurev.es.12.110181.001301. [Classic reference on cannibalism.]

Polski, A. A., Osborn, K. J., Hayward, J. L., Joo, E., Mitchell, A. T., Sandler, A. G., and Henson, S. M. 2021. Egg cannibalism as a foraging tactic by less fit glaucous-winged gulls (*Larus glaucescens*). *Wilson Journal of Ornithology* 133:552–567. DOI: 10.1676/20-00072. [Comprehensive study of the behavioral ecology of egg cannibalism in glaucous-winged gulls. Based in part on the undergraduate Honors theses of Ashley Polski and Karen Osborn. Empirical.]

Sandler, A. G., Megna, L. C., Hayward, J. L., Henson, S. M., Tkachuck, C. M., and Tkachuck, R. D. 2016. Every-other-day clutch-initiation synchrony in ring-billed gulls (*Larus delawarensis*). *Wilson Journal of Ornithology* 128:760–765. DOI: 10.1676/15-121.1. [Shows that egg-laying synchrony occurs in a second species of gulls. Both theoretical and empirical: models connected to data.]

Smith, R. S., Weldon, L. M., Hayward, J. L., and Henson, S. M. 2017. Time lags associated with effects of oceanic conditions on seabird breeding in the Salish Sea region of the northern California Current system. *Marine Ornithology* 45:39-42. [Uses logistic regression and model-selection techniques to determine the time of year SST should be measured in order to best explain reproductive success the following breeding season on Protection Island. Based in part on the undergraduate Honors thesis of Rashida Smith. Both theoretical and empirical: models connected to data.]

Weir, S. K., Henson, S. M., Hayward, J. L., Atkins, G. J., Polski, A. A., Watson, W., and Sandler, A. G. 2020. Every-other-day clutch-initiation synchrony as an adaptive response to egg cannibalism in glaucous-winged gulls (*Larus glaucescens*). *Wilson Journal of Ornithology* 132:575–586. DOI: 10.1676/19-82. [Demonstration that egg-laying synchrony is adaptive in the presence of egg cannibalism in gulls. Based in part on the undergraduate Honors thesis of Sumiko Weir. Both theoretical and empirical: models connected to data.]

V

Appendix

Linear Algebra Basics

This appendix is a self-guided tutorial through the basics of matrices and linear algebra. The student who works through the examples and exercises in this chapter will be prepared for the linear algebra required in this textbook. These exercises should be worked "by hand."

A.1 MATRIX OPERATIONS

A $n \times m$ *matrix* is an array of numbers arranged in n rows and m columns. For example, this is a 2×3 matrix:

$$\begin{pmatrix} 1 & -2 & 4 \\ -1 & 0 & 3 \end{pmatrix}.$$

A.1.1 Matrix Addition

Matrices of the same dimensions can be added. Matrix addition is entry-wise, for example

$$\begin{pmatrix} 1 & -2 \\ -1 & 0 \end{pmatrix} + \begin{pmatrix} 2 & -1 \\ 1 & -3 \end{pmatrix} = \begin{pmatrix} 1+2 & -2+(-1) \\ -1+1 & 0+(-3) \end{pmatrix} = \begin{pmatrix} 3 & -3 \\ 0 & -3 \end{pmatrix}.$$

A.1.2 Scalar Multiplication

A matrix can be multiplied by a *scalar* (real number). Scalar multiplication is entry-wise, for example

$$4 \begin{pmatrix} 1 & -2 \\ -1 & 0 \end{pmatrix} = \begin{pmatrix} 4(1) & 4(-2) \\ 4(-1) & 4(0) \end{pmatrix} = \begin{pmatrix} 4 & -8 \\ -4 & 0 \end{pmatrix}.$$

A.1.3 Matrix Subtraction

Matrix addition and scalar multiplication together allow us to subtract matrices of equal dimension. For example,

$$
\begin{pmatrix} 1 & -2 \\ -1 & 0 \end{pmatrix} - \begin{pmatrix} 2 & -1 \\ 1 & -3 \end{pmatrix} = \begin{pmatrix} 1 & -2 \\ -1 & 0 \end{pmatrix} + (-1)\begin{pmatrix} 2 & -1 \\ 1 & -3 \end{pmatrix}
$$
$$
= \begin{pmatrix} 1 & -2 \\ -1 & 0 \end{pmatrix} + \begin{pmatrix} -2 & 1 \\ -1 & 3 \end{pmatrix}
$$
$$
= \begin{pmatrix} -1 & -1 \\ -2 & 3 \end{pmatrix}.
$$

This is, of course, just entry-wise subtraction; the intermediate steps are henceforth unnecessary.

A.1.4 Matrix Multiplication

The product of a row vector (a $1 \times n$ matrix) and a column vector (an $n \times 1$ matrix) is computed in the following way:

$$
\begin{pmatrix} a & b & c \end{pmatrix} \begin{pmatrix} d \\ e \\ f \end{pmatrix} = ad + be + cf.
$$

For example,

$$
\begin{pmatrix} 1 & 4 \end{pmatrix} \begin{pmatrix} 2 \\ -3 \end{pmatrix} = 1(2) + 4(-3) = -10.
$$

This is called the *dot product* of the two vectors. Note that the dot product gives a scalar, that is, a 1×1 matrix.

In general, two matrices can be multiplied together if the first has dimension $n \times m$ and the second has dimension $m \times p$. The result is an $n \times p$ matrix in which the (i, j)th entry, meaning the entry in the ith row and jth column, is the dot product of the ith row of the first matrix and the jth column of the second matrix. This sounds more difficult than it is; matrix multiplication is easy to carry out once you see the pattern. Here are several examples:

$$\begin{pmatrix} 1 & 2 & 3 \end{pmatrix} \begin{pmatrix} 4 & 7 \\ 5 & 8 \\ 6 & 9 \end{pmatrix} = \begin{pmatrix} 1(4) + 2(5) + 3(6) & 1(7) + 2(8) + 3(9) \end{pmatrix}$$

$$= \begin{pmatrix} 32 & 50 \end{pmatrix};$$

$$\begin{pmatrix} 1 & -2 \\ -1 & 0 \end{pmatrix} \begin{pmatrix} 2 & -1 \\ 1 & -3 \end{pmatrix} = \begin{pmatrix} 1(2) - 2(1) & 1(-1) - 2(-3) \\ -1(2) + 0(1) & -1(-1) + 0(-3) \end{pmatrix}$$

$$= \begin{pmatrix} 0 & 5 \\ -2 & 1 \end{pmatrix};$$

$$\begin{pmatrix} 1 & -2 \\ -1 & 0 \end{pmatrix} \begin{pmatrix} 2 \\ -3 \end{pmatrix} = \begin{pmatrix} 1(2) - 2(-3) \\ -1(2) + 0(-3) \end{pmatrix} = \begin{pmatrix} 8 \\ -2 \end{pmatrix}.$$

Note that *matrix multiplication is not in general commutative. That is, the order in which matrices are multiplied matters.*

A.1.5   Determinants of Square Matrices

The *determinant* of a 2×2 matrix is the scalar number

$$\begin{vmatrix} a & b \\ c & d \end{vmatrix} = ad - bc.$$

This textbook mostly requires computing determinants for 2×2 matrices. If you need to compute the determinant of a higher-dimensional matrix, simply look up the procedure in a more comprehensive discussion of matrices.

Work all of the following exercises, which are designed to lead you step by step through the major concepts you will need from linear algebra. A list of solutions follows the exercises. You will find a summary of the concepts at the end.

A.2   EXERCISES

1. $\begin{pmatrix} 2 & -1 \\ 1 & -3 \end{pmatrix} \begin{pmatrix} 1 & -2 \\ -1 & 0 \end{pmatrix} =$

2. $\begin{pmatrix} 0 & 2 \\ 3 & -2 \end{pmatrix} \begin{pmatrix} 1 & 1 \\ 1 & -3 \end{pmatrix} =$

3. $\begin{pmatrix} 1 & -2 \\ -1 & 0 \end{pmatrix} \begin{pmatrix} 1 & 0 \\ 0 & 1 \end{pmatrix} =$

4. $\begin{pmatrix} 1 & 0 \\ 0 & 1 \end{pmatrix} \begin{pmatrix} 2 & -1 \\ 1 & -3 \end{pmatrix} =$

5. $\begin{pmatrix} 1 & -2 \\ -1 & 0 \end{pmatrix} \begin{pmatrix} x \\ y \end{pmatrix} =$

6. $\begin{pmatrix} 0 & 2t \\ 3t^3 & -2 \end{pmatrix} \begin{pmatrix} a \\ b \end{pmatrix} =$

7. $\begin{pmatrix} 1 & -2 \\ -1 & 0 \end{pmatrix} - \begin{pmatrix} 2 & -1 \\ 1 & -3 \end{pmatrix} =$

8. $3 \begin{pmatrix} 1 & -2 \\ -1 & 0 \end{pmatrix} + \alpha \begin{pmatrix} 2 & -1 \\ 1 & -3 \end{pmatrix} =$

9. $\begin{pmatrix} 1 & -2 \\ -1 & 0 \end{pmatrix} \begin{pmatrix} x \\ y \end{pmatrix} + \begin{pmatrix} 2t \\ -t^3 \end{pmatrix} =$

10. Compute $A - \lambda I$,

    where λ is a scalar and $A = \begin{pmatrix} 1 & -2 \\ -1 & 0 \end{pmatrix}$. Note: The matrix

    $I = \begin{pmatrix} 1 & 0 \\ 0 & 1 \end{pmatrix}$ is called the *identity matrix*.

11. Compute the determinant. $\begin{vmatrix} 1 & -2 \\ -1 & 0 \end{vmatrix} =$

12. $\begin{vmatrix} 5 & -1 \\ 3 & 6 \end{vmatrix} =$

13. $\begin{vmatrix} (2 - \lambda) & 1 \\ -1 & (-1 - \lambda) \end{vmatrix} =$

14. Find the determinant $|A - \lambda I|$,

where λ is a scalar and $A = \begin{pmatrix} 5 & -1 \\ 3 & 6 \end{pmatrix}$. This is called the *characteristic polynomial* of A.

15. Solve the following equation for λ:

$$|A - \lambda I| = 0,$$

where $A = \begin{pmatrix} 5 & -1 \\ 0 & 6 \end{pmatrix}$. The two λ-solutions are called the *eigenvalues* of the matrix A.

16. Solve the following equation for λ:

$$|A - \lambda I| = 0,$$

where $A = \begin{pmatrix} 1 & 2 \\ -2 & 1 \end{pmatrix}$. That is, find the eigenvalues of A. Hint: The eigenvalues in this case are complex numbers.

17. Find the eigenvalues of $A = \begin{pmatrix} -3 & 2 \\ -2 & 2 \end{pmatrix}$.

18. For each of the eigenvalues in Problem 17, find a vector $v = \begin{pmatrix} v_1 \\ v_2 \end{pmatrix}$ such that $Av = \lambda v$; that is, find a vector v such that v satisfies the vector equation $(A - \lambda I)v = \mathbf{0}$, where $\mathbf{0} = \begin{pmatrix} 0 \\ 0 \end{pmatrix}$ is the zero vector. These vectors v are called *eigenvectors*. You may wish to follow the complete solution given below as you work through this procedure for the first time.

19. Find the eigenvalues and associated eigenvectors for the matrix $A = \begin{pmatrix} 1 & 3 \\ 0 & 2 \end{pmatrix}$.

20. Find the eigenvalues and associated eigenvectors for the matrix $A = \begin{pmatrix} 2 & 3 \\ -1 & -2 \end{pmatrix}$.

A.3 SOLUTIONS

Some of the following solutions omit intermediate steps.

1. $\begin{pmatrix} 2 & -1 \\ 1 & -3 \end{pmatrix} \begin{pmatrix} 1 & -2 \\ -1 & 0 \end{pmatrix} = \begin{pmatrix} 3 & -4 \\ 4 & -2 \end{pmatrix}$

2. $\begin{pmatrix} 0 & 2 \\ 3 & -2 \end{pmatrix} \begin{pmatrix} 1 & 1 \\ 1 & -3 \end{pmatrix} = \begin{pmatrix} 2 & -6 \\ 1 & 9 \end{pmatrix}$

3. $\begin{pmatrix} 1 & -2 \\ -1 & 0 \end{pmatrix} \begin{pmatrix} 1 & 0 \\ 0 & 1 \end{pmatrix} = \begin{pmatrix} 1 & -2 \\ -1 & 0 \end{pmatrix}$

4. $\begin{pmatrix} 1 & 0 \\ 0 & 1 \end{pmatrix} \begin{pmatrix} 2 & -1 \\ 1 & -3 \end{pmatrix} = \begin{pmatrix} 2 & -1 \\ 1 & -3 \end{pmatrix}$

5. $\begin{pmatrix} 1 & -2 \\ -1 & 0 \end{pmatrix} \begin{pmatrix} x \\ y \end{pmatrix} = \begin{pmatrix} x - 2y \\ -x \end{pmatrix}$

6. $\begin{pmatrix} 0 & 2t \\ 3t^3 & -2 \end{pmatrix} \begin{pmatrix} a \\ b \end{pmatrix} = \begin{pmatrix} 2bt \\ 3at^3 - 2b \end{pmatrix}$

7. $\begin{pmatrix} 1 & -2 \\ -1 & 0 \end{pmatrix} - \begin{pmatrix} 2 & -1 \\ 1 & -3 \end{pmatrix} = \begin{pmatrix} -1 & -1 \\ -2 & 3 \end{pmatrix}$

8. $3 \begin{pmatrix} 1 & -2 \\ -1 & 0 \end{pmatrix} + \alpha \begin{pmatrix} 2 & -1 \\ 1 & -3 \end{pmatrix} = \begin{pmatrix} 3 + 2\alpha & -6 - \alpha \\ -3 + \alpha & -3\alpha \end{pmatrix}$

9. $\begin{pmatrix} 1 & -2 \\ -1 & 0 \end{pmatrix} \begin{pmatrix} x \\ y \end{pmatrix} + \begin{pmatrix} 2t \\ -t^3 \end{pmatrix} = \begin{pmatrix} 2t + x - 2y \\ -x - t^3 \end{pmatrix}$

10. $A - \lambda I = \begin{pmatrix} 1 & -2 \\ -1 & 0 \end{pmatrix} - \lambda \begin{pmatrix} 1 & 0 \\ 0 & 1 \end{pmatrix} = \begin{pmatrix} (1 - \lambda) & -2 \\ -1 & -\lambda \end{pmatrix}$

11. $\begin{vmatrix} 1 & -2 \\ -1 & 0 \end{vmatrix} = -2$

12. $\begin{vmatrix} 5 & -1 \\ 3 & 6 \end{vmatrix} = 33$

13. $\begin{vmatrix} (2 - \lambda) & 1 \\ -1 & (-1 - \lambda) \end{vmatrix} = \lambda^2 - \lambda - 1$

14. The characteristic polynomial of A is

$$\left| \begin{pmatrix} 5 & -1 \\ 3 & 6 \end{pmatrix} - \lambda \begin{pmatrix} 1 & 0 \\ 0 & 1 \end{pmatrix} \right| = \left| \begin{pmatrix} 5-\lambda & -1 \\ 3 & 6-\lambda \end{pmatrix} \right|$$
$$= \lambda^2 - 11\lambda + 33$$

15. $|A - \lambda I| = 0$ reduces to $(\lambda - 5)(\lambda - 6) = 0$, which holds if and only if $\lambda = 5, 6$. These two numbers are the eigenvalues of A.

16. The quadratic formula leads to the eigenvalues $\lambda = 1 \pm 2i$.

17. The eigenvalues of A are $\lambda = 1, -2$.

18. The eigenvalues are $\lambda = 1, -2$. Also,

$$(A - \lambda I)v = \mathbf{0}$$

is equivalent to

$$\begin{pmatrix} -3-\lambda & 2 \\ -2 & 2-\lambda \end{pmatrix} \begin{pmatrix} v_1 \\ v_2 \end{pmatrix} = \begin{pmatrix} 0 \\ 0 \end{pmatrix},$$

which reduces to

$$\begin{cases} (-3-\lambda)v_1 + 2v_2 = 0 \\ -2v_1 + (2-\lambda)v_2 = 0 \end{cases}. \tag{A.1}$$

For $\lambda = 1$, Equation (A.1) becomes

$$\begin{cases} -4v_1 + 2v_2 = 0 \\ -2v_1 + v_2 = 0 \end{cases}.$$

Thus, we have $v_2 = 2v_1$, so choose $v_1 = 1$ and then $v_2 = 2$. Note that any v_1 and v_2 that satisfy the relationship $v_2 = 2v_1$ will work.

For $\lambda = -2$, equation (A.1) becomes

$$\begin{cases} -v_1 + 2v_2 = 0 \\ -2v_1 + 4v_2 = 0 \end{cases}.$$

Thus, we have $v_2 = \frac{1}{2}v_1$, so choose $v_1 = 2$ and then $v_2 = 1$. Therefore, an eigenvector belonging to eigenvalue $\lambda = 1$ is $v = \begin{pmatrix} 1 \\ 2 \end{pmatrix}$, and an eigenvector belonging to eigenvalue $\lambda = -2$ is $v = \begin{pmatrix} 2 \\ 1 \end{pmatrix}$.

19. The eigenvalues are $\lambda = 1, 2$. An eigenvector belonging to $\lambda = 1$ is $v = \begin{pmatrix} 1 \\ 0 \end{pmatrix}$. An eigenvalue belonging to $\lambda = 2$ is $\begin{pmatrix} 3 \\ 1 \end{pmatrix}$.

20. The eigenvalues are $\lambda = 1, -1$. An eigenvector belonging to $\lambda = 1$ is $v = \begin{pmatrix} -3 \\ 1 \end{pmatrix}$. An eigenvalue belonging to $\lambda = -1$ is $v = \begin{pmatrix} -1 \\ 1 \end{pmatrix}$.

A.4   SUMMARY OF LINEAR ALGEBRA CONCEPTS

Here is a summary of some terms and a note about eigenvectors.

1. The λ-polynomial $|A - \lambda I|$ is called the *characteristic polynomial* of A.

2. The λ-equation $|A - \lambda I| = 0$ is called the *characteristic equation* for A.

3. The λ-solutions of the characteristic equation $|A - \lambda I| = 0$ are called *eigenvalues* of A.

4. The vector equation $(A - \lambda I)v = \mathbf{0}$ is called the *eigenvalue problem* for A.

5. Vector solutions v of the eigenvalue problem $(A - \lambda I)v = \mathbf{0}$ are called *eigenvectors belonging to* λ.

6. The eigenvector belonging to a particular eigenvalue is not unique. For example, if $v = \begin{pmatrix} -1 \\ 1 \end{pmatrix}$ is an eigenvector belonging to λ, then so is $v = \begin{pmatrix} -2 \\ 2 \end{pmatrix}$ and so is $v = \begin{pmatrix} 5 \\ -5 \end{pmatrix}$. In fact, any scalar multiple, that is, any vector of the form $c \begin{pmatrix} -1 \\ 1 \end{pmatrix}$, is an eigenvector.

MATLAB: The Basics

This MATLAB primer is a self-guided tutorial designed to be read while sitting at a computer with MATLAB. The student who reads and successfully works through all the examples in this chapter will obtain the basic working knowledge of MATLAB syntax needed for this textbook.

B.1 PRELIMINARIES

Set the defaults on computer so that file extensions are visible. It is best to leave your computer this way. In Windows 10, right-click on Start > File Explorer > View. Then enable the "File name extensions" by checking the box.

Keep shortcuts for Notepad, Excel, and MATLAB on your desktop or taskbar for easy access. You can find Notepad in Windows 10 by left-clicking on Start.

B.2 SYNTAX AND PROGRAMMING

B.2.1 Command Line

The MATLAB command line prompt is >>. This tutorial will use a single >. You can do simple programming directly at the command line.

Example B.1 *Type the following commands at the prompt. After typing each command, press Enter. This performs the given command and introduces a new prompt below it.*

```
> x = 5
> y = 6
```

B.2.2 Case-Sensitivity

MATLAB is case-sensitive. This includes variables and file names.

B.2.3 Displaying the Current Value of a Variable

Simply type the variable name at the command prompt if you wish to know its current value.

Example B.2 *Type the following commands at the prompt. Here, we assign values to x and y, and then take z to be their sum. If we type z, or x, or z^2 at the prompt, MATLAB prints the value to the screen.*

```
> x = 2
> y = 3
> z = x+y
> z
> x
> z^2
```

B.2.4 Clearing All Variables

At any point, one can clear the memory of all variables by typing
```
> clear
```

B.2.5 Closing MATLAB

To close MATLAB, type
```
> exit
```
or
```
> quit
```
at the command prompt.

B.2.6 Variables and Arithmetic Operators

The equal sign in programming does not mean quite the same thing as it does in mathematics. When you type

> $x = 5$

at the command line in MATLAB, you are saying:

> variable = known numerical value

That is, you are saying that x is the name of a variable and you are assigning the value 5 to it. The right-hand side of an equal sign must be either a number, or a combination of variables that have already been assigned numerical values. For example, in programming, the command > $x = x+1$ is perfectly legitimate if x has already been assigned a value, whereas in algebra, the statement $x = x + 1$ is always a contradiction.

Example B.3 *At the command prompt, type the following commands. In the first line, you are storing the value 5 in the variable x. In the second line, you are storing the value of x (which is now 5 from the previous line) in the variable y, so now y is 5. In the third line, you are storing the value $x+1$ (which is 6) back into the variable x. Now x is 6. Finally, in the last line, you are storing the value of $x+y$ (which is now 11) in the variable z.*

```
> x = 5
> y = x
> x = x + 1
> z = x + y
```

The syntaxes for the five arithmetic operations are

$$+ \quad - \quad * \quad / \quad \char`\^$$

The symbol $\char`\^$ means "raised to the power." The order of operations is the same as it is in arithmetic: compute parentheses, then exponents, then multiplication and division, and then addition and subtraction. When everything is at the same level of operators, the calculation proceeds left to right.

Example B.4 *Type the following commands at the prompt. In the third line, the operation in parentheses is computed first (6+2 is 8), and then that value is squared, which is 64. Then the division is computed (64/2 is 32), and finally, the x is added, giving 32+6, which is 38. The fourth line gives $z = 518$, whereas the fifth line gives $z = 32768$. Make sure you understand the order of operations in each case.*

$> x = 6$
$> y = 2$
$> z = (x + y)^2/y + x$
$> z = (x + y)^(6/y) + x$
$> z = (x + y)^6/(y + x)$

If in doubt about the order of operations, you *can* use extra parentheses to clarify your command, but it is generally bad form. Learn the order of operations.

B.2.7 Programs

Typing a long sequence of complicated commands at the prompt is cumbersome, so we write such sequences of commands in a separate file called a *program* (or *script*, or *code*). When you run the program at the MATLAB prompt, each line within the program is executed sequentially. Here are the general steps for writing and running a program.

1. To write a program using the MATLAB editor, go to "File" in MATLAB; choose "New"and then "Script."

2. Type code (lines of commands) into the file. Do not type > in front of your commands.

3. Save the program file to the folder where you are keeping your MATLAB work. It should be saved as programname.m. Here, "programname" is a placeholder for whatever name you want to call the program. The extension must be *.m for MATLAB.

4. In MATLAB, you need to "set the path,"which means directing MATLAB to the working folder in which you store your programs. The exact way in which you set the path depends on the MATLAB version you are using. You may find "Set Path" under a "File" or "Environment" tab, and many versions have a "browse for folder" button at the top of the Editor window with which you can choose your working folder.

5. To run the program called programname.m, go to the MATLAB command prompt and type the program name without the extension (no ".m").

```
> programname
```

Example B.5 *Type the following three lines in a script file, and save the file to your working folder as hey.m. (The horizontal lines shown here are not part of the program.) To run the program hey.m, simply type > **hey** at the command prompt. The output to the screen will be the values of x, y, and z.*

```
----------
x = 6
y = 2
z = (x + y)^6/y + x
----------
```

B.2.8   Comment Lines

Within a program, the symbol % is used to prevent the execution of a particular line. Commenting out lines is useful for documenting your program and for debugging. For example,

```
%This line is "commented out" and will not execute
```

B.2.9   Printing to the Screen

A semicolon is used at the end of a command if you don't want the value to print to the screen.

Example B.6 *Type the following commands at the command prompt and notice the difference. After you enter the first line and press Enter, the value 5 prints to the screen. In the second line, the value does not print to the screen, although it is still stored in the variable y, as you see after executing the third line.*

```
> x = 5
> y = 6;
> y
```

The semicolon at the end of lines is particularly useful in large programs because printing a lot of output to the screen can drastically slow down the program. Removing the semicolon from selective lines in programs is useful for debugging because it allows you to see the value of a certain variable at a certain point as you run the program.

B.2.10 Numerical Format

The default numerical format for MATLAB is floating-point decimal format. It will report numbers to four decimal places unless the format is changed to "long" in which case it will report to fifteen decimal places. If the format is changed to "rat" the numbers will be reported as ratios (or fractions). One can change the format as follows.

>format long
>format short
>format rat

Scientific notation is denoted by e+###. For example, Avogadro's number could be entered as $6.022e + 23$ or $6.022 \times 10^{\wedge}23$. Either way, it would appear in the short format setting as $6.0220e + 23$.

B.2.11 Loops

Loops are used to repeat sections of the code (program) a specified number of times. The syntax is:

```
for countername = startvalue : increment : endvalue
    (various commands)
end
```

Note that we indent the commands that are inside the loop. This helps highlight the logical structure of the code.

Example B.7 *Write the following code (program). Save it as tschuss.m. Note the comment lines that document the code. Note also that MATLAB does not notice the extra vertical and horizontal spacing. Use such spacing liberally to make your code easy to read. To run the program, type > tschuss at the command prompt.*

```
-------------------------------
%Program tschuss.m
%Initialize x
x = 2

%Loop
for i = 1 : 1 : 6
    x = x * x + i
end
-------------------------------
```

Note that the "end" statement closes the loop.

B.2.12 Crashing a Program on Purpose

If you want to stop the execution of a long program, type control-c.

Example B.8 *Create and run the following program. Call it wiegehts.m. The goal is to print the final value of x to the screen. Suppose that you forget to put a semicolon after the statement x = i+100. All the intermediate values of x will write to the screen, and the program will take a long time to run because of this. While the program is running, crash it using control-c. Now put a semicolon at the end of statement x = i+100 to suppress the intermediate screen output. Save the program file, and rerun the program. Note the speed at which the program runs when it doesn't have to print the intermediate values of x to the screen. Also note that whenever you modify a program file, you must save it before rerunning it.*

```
------------------------------

%Program wiegehts.m
for i = 1 : 1 : 100000
    x = i+100
end
x

------------------------------
```

B.2.13 Logical Statements (If-Then-Else)

The syntax for logical operators is

&	and
|	or
>	greater than
>=	greater than or equal to
<	less than
<=	less than or equal to
==	equal to

Example B.9 *Write and run the following code called hola.m. Try various initial values of x. Don't forget to save the program file after each change before you rerun the program.*

```
------------------------------
%Program hola.m
%Initialize x
x = 7

%Set the variable "check" according to the value of x
if x < 5
    check = 0
elseif x >= 5 & x < 10
    check = 1
else
    check = 2
end
------------------------------
```

Here, the "end" statement closes the logical decision tree.

B.2.14 Input and Output Files

Input files are text (*.txt) files created with Notepad. They are used to import vectors or matrices of numerical data into MATLAB. The input file should be in the working folder where you keep your programs. The command to load the input file is

> `load inputfilename.txt`

The input matrix is saved in a variable called "inputfilename".

Example B.10 *Open a new Notepad file and type the first 10 positive integers in a column vector. Save this input file as ciao.txt. Go to the command prompt and type the following commands. After you load ciao.txt, the vector of numbers from the input file is stored in a variable named "ciao". The second command renames the vector of numbers, calling it "data". The third command adds one to each entry in the vector.*

> `load ciao.txt`
> `data = ciao`
> `data = data + 1`

Example B.11 *Suppose that, after typing the commands in the previous example, you want to save the new values stored in "data" to an output file called "bye.txt". Type the following command. MATLAB will create the file bye.txt in your working folder. Open bye.txt with Notepad and see what it looks like.*

```
>save bye.txt data -ascii
```

In general, the syntax for saving a variable to an output file is
```
>save outputfilename.txt variablename -ascii
```

B.2.15 Creating Functions

Here are the general steps for creating a function in MATLAB.

1. Open a new script file and save it as functionname.m. Here, "functionname" is a placeholder for whatever name you want to give the function.

2. The first executable line in the script should declare that this is a function. The declaration line should look like
```
function y = functionname(inputvariable)
```
3. Define the function in the next lines.

4. Close the function definition with end.

5. At the command prompt, evaluate the function at a particular value by typing
```
> functionname(value)
```

Example B.12 *Create a script file for $f(x) = x^2$ using the following code. Save the program as "square.m". Note that the name of the program file should be the same as in the function declaration line.*

```
-----------------------------
%Create the squaring function square.m
function y = square(x)

    %Define y as a function of x
    y = x^2;

end
-----------------------------
```

After saving the above program to your working folder, type the following commands at the prompt.

```
> square(5)
> square(10)
```

The screen output should be 25 and 100.

B.2.16   Subroutines

A subroutine is a subsidiary program that is called within a main program. It is often easier to see the structure of the main program when certain groups of commands are relegated to subroutines.

Example B.13 *Create a subroutine called sub.m that is called by a program called main.m. You will write two different programs as shown below. Declaring variables as* global *means they are shared between the two programs. With pencil and paper, work through the first several loops of the program and make sure you understand what the program does at each step. Then run the main program at the command prompt by typing* > main.

```
----------
%Program main.m
%Calls subroutine sub.m
%Declare global variables
global x check

%Initialize x
x = 0;

%Loop
for i = 1 : 0.5 : 20
    x = x + i;
    %Call the subroutine sub.m
    sub;
    %Print out value of the variable check
    check
end
----------
```

```
%Subroutine sub.m
%Called by main.m
%Declare global variables
global x check

%Logical decision used to set value of check
if x < 5
    check = 0;
elseif x >= 5 & x < 10
    check = 1;
else
    check = 2;
end
----------
```

B.2.17   Vectors, Matrices, and Arrays

Example B.14 *Type the following commands at the command prompt and note the output. Make sure you understand each operation. The parenthetical explanatory remarks are not part of the commands.*

> x = [1 -3 5 9]
> y = [2 0 3 -1]
> a = 0 : 0.2 : 3
> a = a' (transpose operation)
> n = length(y) (finds the "length" of the vector = number of entries)
> z = x + y (addition of vectors/matrices)
> zz = sum(y) (this is the sum of the entries of the vector y)
> x = x'
> w = [1 -3 4]'
> w(1) (returns the first entry of the vector w)
> w(3) (returns the third entry of the vector w)
> y = [1 3 5; 2 4 -3; 7 -2 0] (semicolons denote different rows)
> y = y'
> x = [1 2 5
3 7 1
-2 0 0] (line breaks denote different rows)

```
> z = x + y
> z = x*y (multiplication of matrices)
> z = x.*y (entry-wise multiplication of matrices)
> z = 2*x (scalar multiplication of matrices)
> z = x/2 (scalar division of matrices)
> z = x.^2 (entry-wise squaring of matrix entries)
> z = x + 1
```

The syntax for locating an entry in a matrix (or an array) is
```
> arrayname(row,column)
```

Example B.15 *Type the following commands at the command prompt.*

```
> x = [1 2 5
3 7 1
-2 0 0]
> z = x(2,3)
> z = x(1,2)
```

A colon can be used to denote "through". When standing alone, the colon denotes all rows or columns.

Example B.16 *Using the same matrix x as in the previous example, type the following commands.*

```
> z = x(2:3,1) (second through third row, first column)
> z = x(:,3) (all rows, third column)
> z = x(1,:) (first row, all columns)
```

One can "stack" vectors or matrices of compatible dimensions, and this can be done horizontally or vertically, as you see in the next example.

Example B.17 *Try the following commands and note the result of each type of syntax.*

```
> y = [1 2]
> v = [y 3]
> w = [y [3 4]]
> z = [y; [3 4]]
```

Example B.18 *There are three commands for building special vectors/matrices. Try these examples at the command prompt.*

```
> z = ones(2,5)  (matrix with 2 rows and 5 columns of ones)
> z = zeros(3,1)  (matrix with 3 rows and 1 column of zeros)
> z = []  (empty matrix/vector)
```

Loops can be used to build vectors, as well.

Example B.19 *Try the following two sample programs.*

```
----------
%Initialize x as a column vector of 10 zeros
x = zeros(10,1);

%Loop
for i = 1 : 1 : 10
    x(i) = 3*i;
end
x
----------
```

An alternative, but slower way to write the above program is

```
----------
%Initialize x as empty vector
x = [];

for i = 1 : 1 : 10
    x = [x; 3*i];
end
x
----------
```

B.2.18 Functions in the MATLAB Library

Many functions already exist in the MATLAB library. The following exercise contains the syntax for some of the most common functions. The student can look up the syntax for other functions in the MATLAB help files as needed.

Example B.20 *Type the following commands at the prompt. Note what each command accomplishes. The parenthetical remarks are not part of the commands.*

```
> x = 2
> y = [1 2 3 4]
> log10(x)  (this is log base 10)
> log10(y)
> log(y)  (this is log base e, the natural log)
> sin(x)
> cos(y)
> tan(y)
> atan(y)  (this is the arctangent of y)
> sec(x)
> csc(x)
> cot(y)
> exp(x)  (this is e raised to the x power)
> exp(y)
> sqrt(x)
```

B.2.19 Plotting

To plot vector y against vector x, use the command
```
> plot(x,y)
```

Example B.21 *Type the following commands at the prompt. Note what each command accomplishes.*

```
> x = 1 : 0.1 : 8
> y = sin(x)
> plot(x,y)
```

B.2.20 Simulating Discrete-Time Models

Discrete-time models, which are simply iterative rules, run quickly on computers.

Example B.22 *Write a program to simulate the Ricker model*

$$x_{t+1} = bx_t e^{-cx_t}$$
$$x_0 = 1.$$

Set $b = 8$ and $c = 0.01$. Find x_{50}.

One way to write this program is to iteratively update the value of x, as in the following code.

```
----------
%Program Ricker.m
%Initialize x and set parameters b and c
x = 1;
c = 0.01;
b = 8;

%Loop
for i = 1 : 1 : 50
    x = b*x*exp(-c*x);
end

%Output value to screen
x
----------
```

A better way to update the value of x is the following.

```
----------
%Program Ricker.m
%Initialize x and set parameters b and c
x = 1;
c = 0.01;
b = 8;

%Loop
for i = 1 : 1 : 50
    xnew = b*x*exp(-c*x);
    x = xnew;
end

%Output value to screen
x
----------
```

An even better way to write this program is to treat x as a vector and define each entry.

```
%Program Ricker.m
%Initialize x and set parameters b and c
x = 1;
c = 0.01;
b = 8;

%Loop
for i = 1 : 1 : 50
    x(i+1) = b*x(i)*exp(-c*x(i));
end

%Output value to screen
x(51)

%plot x vs. t for whole trajectory
time = 1 : 1 : 51;
plot(time,x)
```

In the last program above, make sure you understand why $x(51)$ is x_{50}. Hint: In MATLAB, a vector cannot have a zeroth entry.

Example B.23 *Write a program to simulate this two-dimensional Ricker model.*

$$
\begin{aligned}
x_{t+1} &= by_t e^{-cy_t} \\
y_{t+1} &= x_t + (1-a)y_t \\
x_0 &= 1 \\
y_0 &= 2.
\end{aligned}
$$

Set $a = 0.2$, $b = 8$, and $c = 0.01$. Find x_{50} and y_{50}.

```
%Program Ricker2D.m
%Set parameters
```

```
c = 0.01;
b = 8;
a = 0.2;

%Initialize x and y
x = 1;
y = 2;

%Loop
for i = 1 : 1 : 50
    xnew = b*y*exp(-c*y);
    ynew = x + (1-a)*y;
    x = xnew;
    y = ynew;
end

%Output values to screen
x
y
```

Note that in the two-dimensional Ricker program above, x and y *cannot* be updated with the code

```
x = b*y*exp(-c*y);
y = x + (1-a)*y;
```

because the second line would use the updated value of x on its right-hand side instead of the previous value of x. It is very important to understand this.

Often we want to plot the whole trajectory of x and y. Here is a better way to code the previous example.

```
%Program Ricker2D.m
%Set parameters
c = 0.01;
b = 8;
a = 0.2;
```

```
%Initialize x and y
x = 1;
y = 2;

%Loop
for i = 1 : 1 : 50
    x(i+1) = b*y(i)*exp(-c*y(i));
    y(i+1) = x(i) + (1-a)*y(i);
end

%Output values to screen
x(51)
y(51)

%Plot x and y on the same axes
time = 1 : 1 : 51;
plot(time,x,time,y)
```

B.2.21 Simulating Ordinary Differential Equations (ODEs)

Simulations of continuous-time models such as ODEs tend to run much more slowly than those of discrete-time models. The syntax for the MATLAB *ode45* integrator is

> ode45('RHS', [starttime endtime], initialcondition)

where RHS refers to a program that computes the right-hand side of the differential equation.

Example B.24 *This example concerns a famous model from population dynamics called the logistic model. Integrate the logistic model*

$$\frac{dN}{dt} = 2N \left(1 - \frac{N}{100}\right)$$
$$N(0) = 5$$

from $t = 0$ to $t = 42$. Here, $N(t)$ is the population size at time t. The statement $N(0) = 5$ is called the initial condition. Graph the population size N vs. time t. What is $N(42)$?

To do this example, first write a function program called calcderiv.m that computes the RHS of the logistic ODE.

```
----------
%Program calcderiv.m
function Nprime = calcderiv(t,N)

    %Define RHS of ODE
    Nprime = 2*N*(1-N/100);
end
----------
```

Then, in MATLAB, go to the command prompt and type the following commands.

```
> [T pop] = ode45('calcderiv',[0 42], 5)
> plot(T, pop)
> n = length(T)
> pop(n)
```

Or, write a program called logistic.m that executes these same lines:

```
----------
%Program logistic.m
[T pop]=ode45('calcderiv', [0 42], 5);
plot(T, pop)
n = length(T);
pop(n)
----------
```

and then call the program from MATLAB:

```
> logistic
```

Note that in the example above, T is a vector of times along the trajectory and *pop* is the vector of corresponding N values. The command > n = length(T) sets n to be the number of entries in the vector T. The command > pop(n) picks off the nth value of the vector *pop*, which is the very last value, $N(42)$, which is the population size at time 42. Make sure you understand these variables by typing them at the command prompt to see their values.

```
> T
> pop
```

> n
> pop(1)
> pop(5)
> pop(n)

The last three commands above are not function evaluations. They return the 1st, 5th, and nth entries, respectively, of the vector *pop*.

B.2.22   The Downhill Nelder-Mead Algorithm

The Nelder-Mead algorithm is a "downhill method" of numerically finding minima of functions. MATLAB already has this algorithm in its library of functions. The syntax is

> [x,fval] = fminsearch('functionname',x0)

Here, $x0$ is the initial "guess" from which the algorithm starts downhill, x is the value of the independent variable that minimizes the function, and $fval$ is the function value at the minimum. The independent variable x can be a scalar or a vector.

Example B.25 *Write a program that finds a "minimizer" of the multivariate function $g(a,b) = (a-2)^2 + (b-3)^2 + 5$. That is, write a program that finds values of a and b that minimize the value of $(a-2)^2 + (b-3)^2 + 5$. You will need to write two programs: a main program and a subroutine that computes the value of $g(a,b)$.*

```
%Program findmin.m
%Calls fminsearch.m and gfun.m


%Set initial "guess" for a and b
theta = [0.1 0.03]';
[x fval] = fminsearch('gfun',theta)
```

```
%Program gfun.m
function y = gfun(theta)

    a = theta(1);
    b = theta(2);
    y = (a-2)^2 + (b-3)^2 + 5;
end
```

To run the above program, type
> findmin
at the command prompt. Your output to the screen should be $x =$ $[2,3]'$, and $fval = 5$. Here, x is the minimizer and $fval$ is the value of the function $g(a, b)$ at the minimizer.

B.3 EXERCISES

1. Write a program that finds minimizers of $f(x) = \sin x$. As you know, $\sin x$ has infinitely many minima. Note how your answer depends on your initial "guess" $x0$. Attach your programs and screen output for several values of $x0$.

2. Use the Nelder-Mead algorithm to solve numerically the transcendental equation $x = e^{-x}$. Hint: Let $g(x) = (x - e^{-x})^2$ and find the minimizer for g. Attach your programs and screen output.

3. Write a program to simulate the discrete-time system

$$
\begin{aligned}
x_{t+1} &= \frac{2y_t}{1 + y_t} + 0.25x_t \\
y_{t+1} &= 0.50x_t + 0.7y_t \\
x_0 &= 1 \\
y_0 &= 1
\end{aligned}
$$

from $t = 0$ to $t = 100$. Plot x vs. t and y vs. t on the same axes.

4. Consider the following model of a population of bacteria, where time t is in hours and $N(t)$ is the number of bacteria at time t.

$$
\begin{aligned}
\frac{dN}{dt} &= 0.03\,N\,(N - 15)\left(1 - \frac{N}{100}\right) \\
N(0) &= 17.
\end{aligned}
$$

Numerically integrate this model from $t = 0$ to $t = 1$. What is the value of $N(1)$? Attach your programs and your output.

5. Consider the model in Exercise 4. Write a program that makes a list of the model predictions at the top of each hour. In particular,

use a loop and *ode45* to build a vertical vector x having 20 entries, such that $x(1) = 17$, $x(2) = N(1)$, $x(3) = N(2)$, etc. Save the vector to an output file. Attach your programs and your output file.

Connecting Models to Data: A Brief Summary with Sample Codes

This appendix outlines some methods of connecting models to data and gives samples of MATLAB code that will help students implement the methods on the computer. Students unfamiliar with MATLAB should first work through the self-guided MATLAB tutorial in Appendix B.

C.1 PARAMETERIZATION

The main ideas are:

1. A model, say $N(t) = at^2 + b$, has dependent variables (in this case, N), independent variables (in this case, t), and parameters (in this case, a and b).

2. Imagine graphing the model curve N vs. t as well as observed data points (t, n) as a scatter plot on the same axes. The idea is to make the curve "fit" the data points.

3. We adjust the shape of the curve by "tuning" the values of the parameters a and b.

4. We want to find the values of a and b that make the curve "best" fit the data. Such values $\hat{a}$ and $\hat{b}$ are the parameter estimates. There are various methods for obtaining parameter estimates.

The two methods discussed in this book are maximum likelihood (ML) and least squares (LS).

C.2   RESIDUAL ERRORS (RESIDUALS)

Let's call the model prediction *pred* and the associated data point observation *obs*.

1. The residual error is the observed minus the predicted: $res = obs - pred$.

2. In general, we apply a variance-stabilizing transformation ϕ to make noise approximately additive on the ϕ-scale. Thus, the residual on the ϕ-scale is res = ϕ (obs) $- \phi$ (pred).

3. For demographic noise $\phi(\cdot) = \sqrt{\cdot}$.

4. For environmental noise $\phi(\cdot) = \ln(\cdot)$.

5. For dynamic models such as differential equations and difference equations, it is sometimes best to compute "one-step residuals," in which the model prediction *pred* is conditioned on the observation at the previous time step.

C.3   RSS AND R^2

Suppose noise is approximately additive on the ϕ-scale.

1. The sum of squared residuals (residual sum of squares, or RSS) is

$$\text{RSS} = \sum_{\text{data}} (\phi\,(\text{obs}) - \phi\,(\text{pred}))^2 .$$

2. A generalized goodness of fit is

$$R^2 = 1 - \frac{\sum\limits_{\text{data}} (\phi\,(\text{obs}) - \phi\,(\text{pred}))^2}{\sum\limits_{\text{data}} \left(\phi\,(\text{obs}) - \overline{\phi\,(\text{obs})}\right)^2},$$

where $\overline{\phi\,(\text{obs})}$ denotes the sample mean of the transformed observations.

3. R^2 gives the proportion of variability in the data that is explained by the model, relative to using the mean of the data as a predictor.

C.4 MAXIMUM LIKELIHOOD (ML) PARAMETERS

In this section, we assume data and predictions have already been transformed by ϕ.

1. The residuals depend on the values of the model predictions, which depend on the values of the parameters. So if we "tune" the parameters, we change the values of the residuals.

2. Suppose noise is Gaussian with mean zero and constant variance σ^2 and that stochastic perturbations are uncorrelated in time. The likelihood that a given residual res comes from the normal distribution with mean zero and standard deviation σ is

$$\frac{1}{\sigma\sqrt{2\pi}}e^{-\frac{1}{2}\left(\frac{res}{\sigma}\right)^2}.$$

3. For a model with one dependent variable, the likelihood that ALL the residuals come from the distribution is the product

$$L = \prod_{\text{data}} \frac{1}{\sigma\sqrt{2\pi}}e^{-\frac{1}{2}\left(\frac{res}{\sigma}\right)^2}.$$

4. The likelihood function L is a function of the model parameters.

5. The ML parameter estimates are those parameter values that maximize L, and these are equivalent to those that maximize the log-likelihood $\ln L$.

6. If there are q residuals, the log-likelihood is

$$\begin{aligned}
\ln L &= -q\ln\sigma - \frac{q}{2}\ln(2\pi) - \frac{1}{2\sigma^2}\sum_{\text{data}}(res)^2 \\
&= -q\ln\sigma - \frac{q}{2}\ln(2\pi) - \frac{1}{2\sigma^2}\text{RSS}.
\end{aligned}$$

6. Maximizing the log-likelihood in this context is equivalent to minimizing the RSS.

C.5 LEAST SQUARES (LS) PARAMETERS

Note that:

1. The residuals, and hence RSS, are functions of the parameter values.

2. One can use the Nelder-Mead algorithm, a downhill search method, to find the parameter values that minimize the *RSS* function.

3. The minimizing parameters are the LS parameter estimates.

4. It is helpful to note that the LS method relaxes the restrictive assumptions on the residuals; LS parameter estimates converge to the true values even if the noise is non-normal and autocorrelated, as long as the noise has a stationary distribution.

C.6   IMPLEMENTATION IN CODE

C.6.1   Basic Structure of Program

A code to estimate parameters requires three basic parts:

1. The main program, which sets the initial parameter vector "guess," then passes this initial vector and the name of the RSS function to the downhill search algorithm (which attempts to converge on the minimizers using this initial "guess") and, finally, recovers the parameter estimate vector from the downhill algorithm.

2. The downhill search program. This is not something you need to code yourself; it is likely in the library of functions of the language you are using. For example, in MATLAB the downhill search function is called *fminsearch*. When the main code passes the initial parameter vector and the name of the RSS routine to this function, it uses those initial parameters to compute RSS. Then it "walks downhill" until it finds a local minimum value of RSS. Finally, it returns the minimizer vector of parameters to the main code.

3. The subroutine that computes the RSS as a function of the parameters. This is where the model is coded, predictions and residuals are computed, and the RSS value is computed for that parameter vector.

Note: Most downhill search functions search for negative as well as positive minimizers. In population models, however, parameters are generally positive and we do not want to search negative parameter space. One way to address this is to pass the initial parameter vector v to the downhill search routine as $\ln(v)$ so that it can search all of parameter space. In the RSS routine, the $\ln(v)$ must be converted back via exponentiation before the parameters are used to compute model predictions. Also, when the final parameter estimates are returned to the main program, they must be exponentiated to recover the best parameters on the correct scale.

C.6.2 Constructing Input Files

To construct input files:

1. Prepare an MS Excel data file with the appropriate columns.

2. Copy the contents of the Excel file and paste them into a Notepad file. If you keep the column headings, the headings row must be commented out with an initial %.

3. Save the Notepad file in your program folder as inputname.txt.

C.6.3 Example: Algebraic Model Using Vectors

The goal in this example is to fit the *Gompertz model*

$$f(x) = K \exp\left(-e^{a-bt}\right)$$

to the humerus estimation data in Data Set 2.4 (see Chapter 2) using $\hat{K} = 11.96923077$. Assume environmental noise. We want to find the best-fit parameters a and b.

```
----------
%Program gompertz.m
%Parameterizes the Gompertz model
global age hum


%Load data
load estimation.txt;
data = estimation;
age = data(:,1);
hum = data(:,2);
```

```
%Initialize parameters
theta = log([0.01 0.03]');
```

```
%Run Nelder-Mead algorithm and output best parameters
to screen
[output RSSbest] = fminsearch('RSSgom',theta);
params = exp(output);
a = params(1)
b = params(2)
```

```
%Program RSSgom.m
function RSS = RSSgom(theta)
global age hum
```

```
%Compute vector of predictions
a = exp(theta(1));
b = exp(theta(2));
K = 11.96923077;
pred = K*exp(-exp(a-b*age));
```

```
%Compute RSS on log scale for environmental noise
residuals = log(hum) - log(pred);
squareresiduals = residuals.^2;
RSS = sum(squareresiduals);
end
```

C.6.4   Example: Algebraic Model Using Loop

The goal in this example is to fit the piecewise-linear model

$$f(x) = \begin{cases} a(x-b) + K & x < b \\ K & x \geq b \end{cases}$$

to the humerus estimation data in Data Set 2.4 (see Chapter 2) using $\widehat{K} = 11.96923077$. Assume demographic noise. We want to find the best-fit parameters a and b.

```
----------
%Program lin.m
%Parameterizes the linear model
global age hum


%Load data
load estimation.txt;
data = estimation;
age = data(:,1);
hum = data(:,2);


%Initialize parameters
theta = log([0.1 15]');


%Run Nelder-Mead algorithm and output best parameters
to screen
    [output RSSbest] = fminsearch('RSSlin',theta);
    params = exp(output);
    a = params(1)
    b = params(2)
    ----------
%Program RSSlin.m
function RSS = RSSlin(theta)
global age hum


%Compute vector of predictions
a = exp(theta(1));
b = exp(theta(2));
K = 11.96923077;
n = length(age);

for i = 1 : n
    if age(i) < b
        pred(i) = a*(age(i)-b) + K;
    else
```

```
            pred(i) = K;
        end
    end
    pred = pred';


    %Compute RSS on sqrt scale for demographic noise
    residuals = sqrt(hum) - sqrt(pred);
    squareresiduals = residuals.^2;
    RSS = sum(squareresiduals);

end
```

C.6.5 Example: Scalar Map with One-Step Predictions

The goal is to fit the Ricker model

$$N_{t+1} = aN_t e^{-bN_t}$$

to the data in Data Set C.1. That is, we want to find the best-fit parameters a and b.

```
%Program ricker.m
%Parameterizes the Ricker model
global obst obsN


%load data
load RickerInput.txt;
data = RickerInput;
obst = data(:,1);
obsN = data(:,2);


%Initialize a and b
theta = log([2 0.03]');
```

```
%Run Nelder-Mead algorithm and output best parameters
to screen
[output RSSbest] = fminsearch('RSSfun',theta);
params = exp(output);
a = params(1)
b = params(2)
----------
%Function RSSfun.m
function RSS = RSSfun(theta)
global obst obsN


%Compute vector of one-step predictions
```
$a = \exp(\text{theta}(1));$
$b = \exp(\text{theta}(2));$
$\text{predN} = a * \text{obsN} .*\exp(-b * \text{obsN});$


```
%Compute RSS
numobs = length(obsN);
obsNahead = obsN(2:numobs);
predNlag = predN(1:numobs-1);
residuals = obsNahead - predNlag;
squareresiduals = residuals.^2;
RSS = sum(squareresiduals);

end
----------
```

C.6.6 Example: Higher-Dimensional Discrete-Time Model with One-Step Predictions

Here we fit the environmental noise LPA model to the data in Data Set 6.1.

```
----------
%Program LPA.m
%Parameterizes the LPA model
global Ldata Pdata Adata
```

```
%Load data: user sets this block
load DataSet61.txt;
data = DataSet61;
reps = 4;

%Loop through replicates and construct separate L, P, A
data matrices
Ldata = [];
Pdata = [];
Adata = [];

for i = 1 : reps
    Ldata = [Ldata data(:,(i-1)*3+1)];
    Pdata = [Pdata data(:,(i-1)*3+2)];
    Adata = [Adata data(:,(i-1)*3+3)];
end

%Initialize parameters
theta = log([
    8
    0.01
    0.01
    0.2
    0.01
    0.2
]);

[output RSSbest] = fminsearch('RSSlpa',theta);
params = exp(output)
----------
%Program RSSlpa.m
function RSS = RSSlpa(theta)
global Ldata Pdata Adata

%Set parameters
par = exp(theta);
b = par(1);
cel = par(2);
cea = par(3);
```

```
mul = par(4);
cpa = par(5);
mua = par(6);

n = length(Ldata(:,1));

%Compute one-step prediction and residual matrices
Lpred = b*Adata.*exp(-cel*Ldata-cea*Adata);
Ppred = (1-mul)*Ldata;
Apred = Pdata.*exp(-cpa*Adata) + (1-mua)*Adata;

Lp = Lpred(1:n-1,:);
Pp = Ppred(1:n-1,:);
Ap = Apred(1:n-1,:);

Lo = Ldata(2:n,:);
Po = Pdata(2:n,:);
Ao = Adata(2:n,:);

Lres = log(Lo) - log(Lp);
Pres = log(Po) - log(Pp);
Ares = log(Ao) - log(Ap);

RSSL = sum(sum(Lres.^2));
RSSP = sum(sum(Pres.^2));
RSSA = sum(sum(Ares.^2));

RSS = RSSL + RSSP + RSSA;

end
```

Index

abiotic factor, 11
Akaike Information Criterion,
    8, 43, 122, 221, 253
Akaike weight, 253
Allee effect, 164, 181
Allee threshold, 165
Ansatz, 93, 189
art of modeling, 6, 114
assumptions
    deterministic, 33
    model, 32
    stochastic, 33
Atkins, G. J., 256
attractor, 77, 79
autonomous, 159, 187, 202
axiom, 20

Beetle Team, 111, 125, 216
Beverton-Holt model, 73,
    82, 156
bifurcation, 78, 202, 203
    blue-sky, 184
    diagram, 61, 76, 79,
        81, 160
    Hopf, 211
    hysteresis, 178, 179
    hysteresis loop, 179
    parameter, 61
    pitchfork, 177
    point, 61
    saddle-node, 177
    transcritical, 176
    vertical, 160
binning, 36

biology, 9
birth rate
    per capita, 57

cannibalism, 113, 247, 249, 256
    flour beetles, 113
    gulls, 247
carrying capacity, 163
Caswell, H., 86, 109
causation
    proximate, 12
    ultimate, 12
    vs. correlation, 12, 241
center, 193, 195
chaos, 79, 125
characteristic equation, 97, 98,
    190, 272
characteristic polynomial,
    269, 272
climate change, 247, 256
closed-form solution, 58
community, 11
compartmental model, 56, 146
    continuous-time, 146
    discrete-time, 56
competition, 86, 205
computer
    code, 41
    program, 41
conditioned least squares, 119,
    120, 220
confidence interval, 240, 242
consistent
    logically, 20

continuously differentiable, 67,
        103, 172, 202
contradiction
        logical, 20, 27
cooperation, 86, 205
correlation
        vs. causation, 12, 241
Costantino, R. F., 111, 124,
        125, 127, 128
cross-sectional study, 31
crowding effects, 60, 162
Cushing, J. M., 111, 114, 124,
        159, 166, 168, 211
cycle chain, 202

Darling, Fraser, 256
data
        estimation, 36
        validation, 36
day of year, 216
deduction, 19
Dennis, Brian, 111
density dependence, 162
derivative, 13
        partial, 85, 187
Desharnais, R. A., 111, 125
determinant, 267
        2×2, 93
        $n \times n$, 98
deterministic, 13, 79
        skeleton, 34, 117
        unpredictability, 80
diagram
        bifurcation, 61, 76, 79, 160
        Leslie, 87, 90, 116
Dickens, Charles, 169
difference equation, 4, 55, 85
        linear scalar, 57
        linear systems, 90
        nonlinear scalar, 60

        nonlinear systems, 99
differential equations, 4, 145,
        159, 187
        linear scalar, 159
        linear systems, 188
        nonlinear scalar, 162
        nonlinear systems, 198
        separable, 151
dimension, 86
displacement from equilibrium,
        66, 101, 182, 199
distribution, 17
        normal, 17
        stable age, 99, 106
Doomsday Model, 166
dot product, 266
doubling time, 71, 166
downhill method, 40
dynamical system, 4
dynamics
        steady state, 226
        transient, 226

ecology, 10
        behavioral, 12
        community, 11
        ecosystem, 11
        mathematical, 12
        physiological, 11
        population, 11
ecosystem, 11
eigensolution, 94, 190
eigenvalue, 59, 65, 66, 94, 97,
        98, 172, 190, 269,
        271, 272
        complex, 194
        dominant, 99, 105
        imaginary, 194
eigenvalue problem, 272

eigenvector, 94, 97, 98, 190, 269, 272
eight-cycle, 78
El Nino-Southern Oscillation, 250
emergent properties, 10
endemic disease, 150
epidemic, 148
epidemiology, 9
epistemology, 22
equation
    autonomous, 159, 187, 202
    difference, 4
    differential, 4, 145
    equilibrium, 14, 92
    fixed-point, 14, 92
    Van der Pol, 204
equilibrium, 14, 15, 59, 64, 92
    asymptotically stable, 15
    hyperbolic, 67, 103, 171, 201
    neutrally stable, 15
    stable, 15
    unstable, 15
error
    measurement, 16, 32
    observational, 16
    process, 16, 32
    residual, 35
Euler's formula, 195
exchange of stability, 176
expected value, 18
exponential growth, 24

factor, 237
    abiotic, 11
fecundity, 57, 90
final state, 79
Fisher, Ronald, 10
fixed point, 14, 92

equation, 14
flour beetle, 23, 86, 99, 111, 112
four-cycle, 78
fractal, 83
Fraser Darling effect, 256
function, 13
    probability density, 17
Fundamental Theorem of Demography, 98

Galusha, J. G., 215
Gaussian
    noise, 117
    variable, 40, 44, 50, 229, 239, 297
general solution, 153
generalized linear model, 242
geometric growth, 58
glaucous-winged gull, 30, 216, 248
Gleick, James, 75
Gompertz model, 299
goodness-of-fit, 8, 45, 124, 296
growth, 30
    bone, 29
    exponential, 24
    geometric, 58
    Malthusian, 24, 57, 58, 161
growth rate
    intrinsic, 59
    per capita, 16, 164
    population, 16
gull
    glaucous-winged, 30, 216, 248

Haldane, J. B. S., 10
half-saturation constant, 47
Hardy, G. H., 10

harmonic oscillator, 193
Hartman-Grobman
       Theorem, 202
Hastings, A., 124
heteroclinic cycle, 202
histogram, 19
Holling Type III, 47
Hopf bifurcation, 211
humanities, 21
Hunt for Chaos
       experiments, 125
hyperbolic, 67, 103, 171, 201
hypotheses
    alternative, 7, 220
hypothesis, 7
hysteresis, 178, 179
hysteresis loop, 179

induction, 19
    mathematical, 20, 26
infectives, 147
information theory, 8, 43
initial condition, 58, 153
    sensitivity to, 79
initial value, 58
initial value problem, 58
interaction
    competition, 86
    cooperation, 86
    interspecific, 86
    intraspecific, 86
    mutualism, 86
    predator-prey, 86
interaction terms
       (regression), 242
intrinsic growth rate, 59, 163
isocline, 204

Jacobian matrix, 102, 199
Janoschek curve, 34

King, A. A., 111
KISS principle, 6, 114

Larus glaucescens, 30, 216, 248
least squares, 297
    conditioned, 120, 121
    nonlinear, 39
Leslie
    diagram, 87, 90, 116
    matrix, 89, 91
    model, 86
    model, nonlinear, 99, 116
Leslie, P. H., 86
Levin, S. A., 28, 228
life-cycle stage, 86
likelihood function, 39, 123,
       297
limit cycle, 202
linearization
    importance of, 63
    of functions, 62, 63
    of scalar difference
       equations, 64, 66
    of scalar differential
       equations, 170
    of systems of difference
       equations, 100, 101
    of systems of differential
       equations, 201
Linearization Theorem, 67, 76,
       103, 172, 202
link function, 242
loafing, 216
log odds, 240
log scale, 35
log-likelihood, 39
log-likelihood function, 297
logistic map, 73, 82
logistic model, 14, 24, 150, 156,
       163, 172, 291

logistic model, (*cont.*)
    discrete-time, 155
logistic regression, 247
longitudinal study, 31
Lotka, Alfred J., 23
Lotka-Volterra model, 204
    competition, 205
    cooperation, 205
    predator-prey, 206
LPA model, 23, 99, 116, 128
    demographic noise, 117
    environmental noise, 117

MacArthur, Robert H., 228
Malthus, Thomas, 161
Malthusian growth, 24, 57, 58, 161
manifold, 191
    stable, 192
    unstable, 191
map, 4, 55
    linear, 57
    nonlinear, 60
    linearization, 64
    logistic, 73, 82, 155
    Ricker, 67, 75
mathematical induction, 20, 26
mathematics
    pure, 20
Matlab
    arithmetic operations, 275
    case sensitivity, 274
    clearing variables, 274
    closing, 274
    code, 276
    command line, 273
    comment lines, 277
    crashing a program, 279
    discrete-time model, 286

    display current value of variable, 274
    fminsearch, 221, 292
    functions, 281, 285
    glmfit, 240, 257, 258
    input files, 280
    logical statements, 279
    logistic model, 290
    loops, 278
    matrices, 283
    minimizer, 292
    Nelder-Mead algorithm, 292
    numerical format, 278
    ode45, 221, 290
    ODEs, 290
    output files, 280
    plot command, 286
    printing to screen, 277
    programs, 276
    Ricker model, 286
    sample codes, 295
    script, 276
    spline, 229
    subroutines, 282
    variables, 274
    vectors, 283
matrices
    noncommutativity of, 267
matrix, 265
    addition, 265
    determinant, 267
    identity, 268
    Jacobian, 102, 199
    Leslie, 89, 91
    multiplication, 266
    multiplication by scalar, 265
    projection, 89
    subtraction, 266

maximizer, 39
maximum likelihood, 38, 296
    parameters, 39
May's Hypothesis, 81
May, Robert, 81, 125, 206
mean, 19
mechanism
    deterministic, 33
    stochastic, 33
metamorphosis, 86, 112
minimizer, 40
model
    fitting, 7, 8, 36, 295
    parameterization, 119, 220
    predictions, 8, 222
    selection, 8, 42, 122, 221
    validation, 8, 36, 45, 124,
        222
    validation in logistic
        regression, 253
model
    age-structured, 86
    algebraic, 4
    assumptions, 6
    Beverton-Holt, 73, 82, 156
    compartmental, 56, 146,
        150, 218
    continuous-time, 4
    demographic noise
        LPA, 119
    deterministic, 33
    difference equation, 4
    differential equation, 4
    discrete logistic, 82
    discrete-time, 4, 85
    discrete-time logistic, 155
    environmental noise
        LPA, 118
    generalized linear, 242
    Gompertz, 299

Janoschek, 34
Leslie, 86
linear, 89
logistic, 14, 24, 150, 156,
    163, 172, 291
Lotka-Volterra, 204
LPA, 23, 99, 116, 128
mathematical, 4
mechanistic, 6
NLAR, 117
nonlinear, 89
nonlinear Leslie, 99
regression, 237
Ricker, 24, 61, 75, 82,
    83, 135
SIR, 147
stage-structured, 86
stochastic, 34, 35
model-averaged parameter
    estimates, 253
modeling
    art of, 6, 114
    ecological, 10
multiple time scale
    analysis, 226
mutualism, 86

natural selection, 12
Nelder-Mead algorithm, 40,
    300
net reproductive number, 107
    inherent, 131
neutrally stable, 15, 195
NLAR model, 117
NOAA, 218
node
    asymptotically stable,
        191, 192
    unstable, 189, 191
noise, 6, 16, 114

noise, (*cont.*)
  demographic, 16, 296
  environmental, 17, 296
  Gaussian, 117, 239, 297
nonlinear autoregressive
      model, 117
nonlinear least squares, 39
normal distribution, 17
nullcline, 204

Occam's Razor, 27
odds, 240
odds ratio, 241
ODE
  linear system, 188
  nonlinear system, 198
orbit, 13
organism, 10
  bones, 30
  development, 30
  growth, 30
  organs, 30
  shape, 30
  size, 30
outbreak, 149

paradox, 27
parameter, 6, 295
  bifurcation, 61
  estimation, 7, 36, 39, 295
  least squares, 39, 296, 297
  maximum likelihood, 39,
      296, 297
  model-averaged, 253
  space, 40
parameterization, 7, 36, 295
partial fractions, 151
per capita growth rate, 164
period-doubling cascade, 79
phase line portrait, 170

phase plane, 190
phase plane portrait
  asymptotically stable
      node, 191
  asymptotically stable
      spiral, 196
  center, 193
  saddle, 192
  six generic, 197
  unstable node, 189
  unstable spiral, 195
phase space, 13
Poincare-Bendixson
      Theorem, 204
population, 11
predator-prey, 86, 206
predictability, 80
prediction, 21
  one-step, 220, 227
probability density function, 17
projection matrix, 89
proportional to, 15
  inversely, 15
Protection Island, 30, 208,
      216, 252
proximate cause, 12

R-squared, 45, 124, 221, 296
random variable
  continuous, 19
  discrete, 19
  realization, 35
  standard normal, 35,
      117, 239
rate
  vital, 86
rate of change, 13
realization, 17
recovereds, 147
recruitment, 60

recursion formula, 4, 72
regression
    coefficients, 238
    gamma, 242
    interaction terms, 242
    intercept, 238
    linear, 237, 242
    logistic, 240, 242, 247, 253
    multiple, 238
    Poisson, 242
    slope, 238
residual, 35, 119, 221, 296
    log, 119
    one-step, 119, 221
    square-root, 35, 119, 120
residual sum of squares, 39,
        120, 221, 296
Ricker model, 24, 61, 67, 75,
        82, 83, 135, 287
Ricker nonlinearity, 70, 89, 131
RSS, 39, 120, 221, 296

saddle, 192, 193
sampling
    stratified random, 36
saturation, 33
scale, 9, 10, 227, 228
    community, 11
    ecosystem, 11
    organism, 10
    population, 11
    species, 11
science, 20
sea surface temperature, 250
seabird behavior, 215
Seabird Ecology Team, 215,
        228, 247
sensitivity to initial
        conditions, 79
sigmoidal growth, 33

signal, 16
sink, 170
SIR model, 147
solar elevation, 216, 217
solution
    closed-form, 58
    general, 95, 97, 98, 153,
        160
    particular, 95, 97, 98, 153,
        160
    singular, 153
source, 170
species, 11
    social, 10
    solitary, 10
spiral
    asymptotically stable,
        196, 197
    unstable, 195, 196
spline, 221
square root scale, 35
stable, 15
    asymptotically, 15
    neutrally, 15
stable age distribution, 99
stage-structured model, 86
standard deviation, 18
standard error, 240
standard normal random
        variable, 35
state space, 13
steady state, 226
stochasticity, 13, 16, 80
    demographic, 16, 17,
        35, 117
    environmental, 16, 17,
        35, 117
    Gaussian, 117
stratified random sampling, 36

Strong Ergodic Property, 99, 105
survivorship, 60
susceptibles, 147
synchrony, 256

tangent plane, 100, 199
Tenebrionidae, 111
thermocline, 250
tide, 217
    semi-diurnal, 217
    height, 216
time series, 13, 76
transformation
    variance-stabilizing, 35, 117, 296
transient, 82, 227
trend
    deterministic, 16
Tribolium, 23, 86, 111
    adult, 112
    castaneum, 99, 111
    egg, 112
    larva, 112
    pupa, 112
two-cycle, 78

ultimate cause, 12
unpredictability, 80
unstable, 15

Van der Pol equation, 204, 210
variable
    binary, 240
    continuous random, 19
    coupled, 13
    dependent, 5, 295
    discrete random, 19
    explanatory, 237
    extrinsic, 6
    independent, 5, 295
    intrinsic, 6
    predictor, 237
    random, 17
    realization of random, 17
    response, 237
    standard normal random, 117, 239
    state, 5
variance, 19, 40
variance-stabilizing
        transformation, 35, 117, 296
variation equation, 66, 182
vital rates, 86

Weldon, L. M., 247
Wright, Sewell, 10

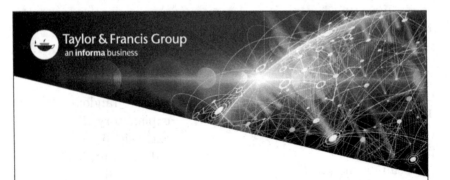

Printed in the United States
by Baker & Taylor Publisher Services